With the Land

*Reflections on Land Work and
Ten Years of the Landworkers' Alliance*

With the Land

Published by Landworkers' Alliance
LANDWORKERSALLIANCE.ORG.UK

Copyright 2023 by Landworkers' Alliance
Printed in the UK by Pureprint Group
ISBN 978-1-3999-5381-8

With the Land has been made possible with the generous support of
The A Team Foundation, Be The Earth and The Roddick Foundation.

This book contains environmentally responsible FSC® certified papers. Cover: 5% hemp fibre, 40% recycled material, and 55% pure environmentally friendly fibre. Inner: 100% post-consumer waste.

This book is a collaborative effort, guided by the vision and experience of many people.

Working Group
Yali Banton-Heath
Joanna Blundell
Kate Briton
Sally Lai
Oli Rodker
Lauren Simpson

Coordinator
Sally Lai

Editorial Advisor
Charlotte Du Cann

Proofreading & Editing
Lucy Badrocke

Editing
Yali Banton-Heath
Joanna Blundell
Sally Lai

Cover Image & Illustrations
Rosanna Morris

Design
Minute Works

Contents

To the Reader

With the Land came into being as we were thinking of ways to mark the first ten years of the Landworkers' Alliance.

Rather than documenting the last ten years in a linear and historical way (another project for another time!), we wanted this to be a book that holds the past, present, and future.

Within these pages, you will find insights into the wider socio-political context and the urgency that brought a group of people together to form the Landworkers' Alliance. You will glean what the alliance means to people today and hear about the work that is left to do, which continues to inspire more and more people to join our growing movement.

As a member-led organisation, it was important to us that this book encompasses a multiplicity of voices and reflects the diversity of our membership. Through our open call for submissions and invitations to contribute, we assembled a collection of contributions that span different regions and sectors; which explore what it means to work the land and, more importantly, to work *with the land*. As individual pieces they highlight some of the important and intersecting issues faced by land-based workers, including access to

land, the importance of seed sovereignty, gender and landwork, and the ecological and social imperative of creating a better food and landwork system by working in harmony with nature. Together, they celebrate the possibility of forging positive change in our food and land-use system through collective action; a vision which lies at the heart of our movement.

We worked with writer and editor Charlotte Du Cann to curate one possible sequence and journey through which to explore the material, but we invite you to experience the contents as you choose, in any order that makes sense to you. Most importantly, we hope that in these pages, filled with song lyrics, letters, poems, texts, essays, and imagery; you will find accounts that resonate, move, inform, and inspire you. As Oli Rodker says in his piece, 'there is a lot more history to go' and more work needed. *With the Land* is an acknowledgement of the work that has gone before and an invitation to join us in the collective endeavour towards food sovereignty, agroecology, and land justice.

Together we can make hope possible.

In solidarity,
With the Land Working Group

EDWIN BROOKS

The Wey and The Lod

The sky forgot the sun today and drowned
the breathless fields; the world a backshot wheel
that spun the birds back to their nests and filled
the muddy thicket up with steaming cows.
The wetted hens all crowd like shipwrecked souls
among the choking grass and tangled weeds.
Now nothing can recall how dryness feels —
the soil is thick with wet it cannot hold.
But just beyond the heath there is a field
that falls across a ridge where my friend said
the watershed runs northwards to the Wey
and southwards to the Lod across the Wield
so tears dashed out on either cheek will slip
to the North Sea. Or to the Atlantic.

Beginnings

Founding member of the Landworkers' Alliance
Ed Hamer sets the scene for its emergence

The emergence of the Landworkers' Alliance is a story of succession. Although there was a defining moment when a handful of aspiring peasants came together to take the first steps on our journey as a union, the roots of the LWA stretch back much further than our collective ten-year history. Our founding drew inspiration from generations of farmers, activists, writers, and campaigners who had been ploughing the same furrow for the best part of half a century.

These are the weathered hands from which we took the baton: from the radical wings of the Soil Association and the West Wales homesteaders in the 60s and 70s, to the founding of the Family Farmers' Association in the 80s; the audacity of The Land Is Ours, born out of the road protest camps and anti-GM actions of the 90s, to those fighting the corner of land-based livelihoods at the anti-globalisation protests of the early 2000s. These are the shoulders on which we stand.

Today the LWA firmly flies the flag of intergenerational governance, but it is fair to say that we rode to prominence on a wave of Millennial conviction. Throughout the 2000s, there was a growing interest and demand for land-based livelihoods as a positive response to the seemingly daunting challenges of climate change, globalisation and later on, austerity; coupled with growing enthusiasm from consumers for high quality, seasonal and local food. There was a clear opportunity for a generation of students, graduates, and unemployed youngsters to skill themselves up in producing food in an effort to be the change they wanted to see in the world. A number of urban growing initiatives, OrganicLea and Grow Heathrow amongst others, inspired many who couldn't readily access land based projects, that they too could dream of a life on the land.

Inevitably, this momentum grew strongest in those towns and cities with close-knit communities of ecological Millennials such as London, Bristol, and Brighton. However, something was also emerging in the countryside. Decades of gentrification and restrictive planning policies had made it near-impossible for a generation of rural youngsters to get onto the farming ladder, or to live on the land if they were fortunate enough to work it. All were convinced in the potential of good food and good farming to change the world for the better, and both were struggling to raise their voices and their profile.

A similar momentum was building in countries across Europe, strengthened by continental affinity. A few of us were involved in the early days of Reclaim the Fields, an alternative young farmers' network that brought together experienced practical landworkers with largely-urban political activists in a common cause. Successive European summits for the G8, World Bank, and UN Climate Conventions provided a focus for progressive farming blocs to share experiences, forge bonds, and reinforce our collective struggle.

This cross-pollination inspired a common cause and also highlighted some clear failings. Despite the UK's respected history of radical direct action, our alternative farming community didn't have the skills to communicate effectively with regional and national policy makers. Our initial forays into lobbying The Department for Environment, Food and Rural Affairs (Defra) were hindered by a lack of identity and cohesive messaging, and were met with an opaque wall of silence or, at best, a suggestion that we raise our concerns with the National Farmers Union (NFU).

We were acutely aware that none of the established farming unions spoke our language or represented anything like the type of farming we embodied: ecological, resilient, labour-led production for local and regional markets. There were of course organisations we respected and with whom we identified, such as The Soil Association, Sustain, the UK Food Group, and Friends of the Earth. They were all making invaluable contributions to food policy discourse, but were all campaigning non-governmental organisations and none of them were led by producers.

Then we discovered the strategic role played by a bold, ambitious and colourful union representing the rights of farmers like us across Europe, America and the Global South called La Via Campesina (LVC). As an umbrella organisation LVC represented over 180 farming unions across 81 countries. Their name and logo were both synonymous with radical campaigns against land-grabbing, commodification, and neoliberal agribusiness; from Santiago in the south to Stockholm in the north.

In 1996, LVC was responsible for founding the concept of Food Sovereignty — the right of communities to define their own culturally appropriate food and farming systems. Collectively they held more than two decades of experience in uniting and mobilising dispersed rural and urban farming communities under a common banner: 'Globalise the struggle, Globalise Hope!'

However, it was not until we started building alliances with farming unions across the channel; Confédération Paysanne in France, Euskal Herriko Nekazarien Elkartasuna / Unión de Agricultores y Ganaderos Vascos (EHNE / UGAV) in the Basque region, and Coordinadora de Organizaciones de Agricultores y Ganaderos (COAG) in Spain, that we recognised the importance of LVC's democratic structures and regional assemblies. Strength and conviction in their identity allowed them to decentralise decision making, while at the same time presenting a united front to international institutions like the UN, to whom they'd become a statutory consultee.

Within Europe, the European Coordination Via Campesina (ECVC) had spent years building coalitions between diverse farming unions and learning how to effectively engage with the civil society mechanisms of the European Commission. On the ground, they were also instrumental in organising demonstrations, marches, skills exchanges, and social events aimed at delivering change and building solidarity.

The founding principle of LVC, that it is exclusively a producer-led organisation, has been fundamental to the success of its lobbying work. Compared to numerous other NGOs campaigning and lobbying on similar topics, LVC and its affiliated unions are respected for bringing first-hand experience from the fields, forests, and fishing grounds to the negotiating table. This gives them a unique role in officially representing marginalised voices within the policy framework.

Back in 2012, the LWA's founding members were a handful of idealistic first and second-generation farmers with big ideas and even bigger ambitions. We were, each of us, playing a part in building the momentum, and knew only too well the barriers facing those like us looking to secure a livelihood on the land. We had encountered LVC's work, either in person or anecdotally, and knew without doubt that they represented not only our destination, but the tools for how to get there.

Ten years ago I was invited by Jyoti Fernandes, now LWA's Policy and Campaigns Coordinator, to a brainstorming session to help launch a UK affiliation of LVC. A handful of us met on a dark, rainy Friday at the end of November 2012 at the Soil Association offices in Bristol, and spent the day appraising the successes of LVC in other European countries, while lamenting the lack of traction of any small farming unions here in the UK.

We must have been inspired with purpose, as I remember towards the end of the day we voted in a chair, accountant, and secretary for our fledgling union. In what would become time-honoured fashion, we then retired around the corner to The Bull on Hillgrove Street to toast the future of radical agrarian political change.

Looking back, we could not have known that bringing together the momentum and the framework would have such a catalysing effect on the active and aspiring farming communities we hoped to serve.

As we celebrate our tenth anniversary as a union, we honour the fact that small actions can create lasting ripples. Along with so many achievements, from our first Farmhack to The Landskills Fair, from the emergence of the LWA youth wing, FLAME, to Land In Our Names (LION), our collective identity has given a new voice to a generation with their hands on the land.

Long may we continue, and in the words of La Via Campesina: *Viva!*

Sowing Seeds

TONI LÖTTER

Tanka Seeds

The seeds you hold, cupped
against the lines of your palm
carry the Earth's plans.
So when you plant them, be still,
for you are guiding magic.

ASHLEY WHEELER

The Beginnings of the UK Seed Sovereignty Movement

Organic vegetable grower and seed enthusiast Ashley Wheeler from Trill Farm Garden looks back at the beginnings of the UK seed sovereignty movement

At the first Landworkers' Alliance meeting in 2012, it was decided that one of the main campaign strands to focus on would be seeds. If truth be told, at that time I was not really sure what the issues around seed were. I was aware of some big multinational companies controlling a lot of the seed across the world, but I did not really know the sinister side of how that came to be — how an invaluable resource that farmers and growers have had full control of for thousands of years was seized from their hands. I was wholly unaware of the tactics these companies used to persuade farmers that their own seed was inferior to the new hybrid seed that they could offer them, or about how they patented indigenous seeds.

Just over a year after the Landworkers' Alliance was formed, three of us headed to Brussels. Legislation around seed in the EU was being discussed, and proposals were being drafted that would inflict tighter controls on farmers' rights to save seed, especially of open pollinated varieties. We went to meet other members of the European Coordination Via Campesina (ECVC), many of whom had been working on seed legislation for years, campaigning for farmers' rights

to grow and save their own seed. We had heard of their experiences, their struggles, and the strategies that they were using to maintain and improve peasant-held seed.

It was at this moment that I felt part of something big, something so much bigger than just us on the small area of land that we had come to look after. We were part of an international movement, working alongside millions of other small scale farmers and growers across the world who were fighting for their rights to produce good, healthy food for their communities in an ecological way. It is easy to feel isolated as a farmer or grower — we live rurally, and the work can often be solitary, but this gathering in Brussels made me realise that we were not alone. It also gave me a heightened sense of purpose as a grower (not that growing food is not enough!). I became impassioned about breaking through the rhetoric that seed production should be left for the experts, and felt strongly that we should be producing more of our own seed and setting up systems to build more resilience, and bring more power to the farmers and growers who use the seed.

At Trill, we began to save more seed — some for ourselves, some to exchange with others, and some for small UK seed companies like Real Seeds, who have been working on this on all of our behalf for years. The three of us who went to Brussels set up the South West Seed Savers Co-op. I became involved in the Gaia Foundation's Seed Sovereignty Programme, and over the last few years I've seen a huge rise in interest in seed sovereignty. We are beginning to reclaim the skills that were once a central part of what it meant to be a farmer or grower, but which have been eroded and left to the 'experts'. It is our belief that we must work to reclaim these skills and increase the availability of healthy and resilient open-pollinated seeds that are appropriate for our growing methods. We must also build systems and communities to bring back the joy that comes with producing seed; and to trial, breed, and exchange seed amongst ourselves. This is something that is now starting to gain momentum and I am proud to be part of the growing seed sovereignty movement now taking root in the UK.

IONE MARIA ROJAS

The Smallest Giver of Life

HUAHTLI, N. FROM NAUHATL

Artist and grower Ione Maria Rojas shares how her relationship with amaranth has shaped her life

> How do you go about finding that thing,
> the nature of which is totally unknown to you?
> — Rebecca Solnit[1]

For me, it began with amaranth. A towering explosion of magenta and gold, I first met amaranth when it was just a bag of seeds, and I was beginning my training in Sustainable Horticulture at Schumacher College. Having been asked to sow the seeds in the circular Hazelip garden, three of us broadcast it over two small beds on the first sunny days of May and waited like nervous parents for it to sprout and grow. Eventually, small, neat leaves emerged out of copper stems, a tie-dye effect across the bed as they unfurled in greens, pinks, purples, and ambers. I was instantly enamoured.

Our plants grew quickly, and thickly, and strong. They grew steadily upwards, as straight as corn or wheat, but with a bolder, louder palette. I watched with glee as in late August they started to flower, long grain-like buds, hot spiky fingers, tentacled sea creatures of seed and rust. A shock of colour in a Devonshire garden. Colours of brightly painted houses in a loud chaotic city, of tissue paper piñatas suspended over market stalls, of *alebrijes*, and bougainvillea, and the sunset after a mountain storm.[2] Watching the amaranth grow transported me to somewhere familiar and forgotten. Fragments of childhood, a family across an ocean, a sleeping dragon lazily blinked one eye open.

I remember when I first read of amaranth's importance as a food staple of the Mexica, I remember the tiny *of course* feeling in my belly. Known as *huauhtli* (the name amaranth arrived with Spanish colonisation), the leaves were eaten as greens and the seeds eaten whole or ground into flour for tortillas and tamales, using the same methods that are used by many Mexicans today. In those early google binges, I found the Spanish codices and Nahua illustrations that depict the *huauhtli* seed tributes paid to the Mexica capital Tenochtitlan by

its provinces. A seed tax. I read descriptions of the offerings made for ceremonies, figurines of the gods made from a *huauhtli* seed and honey dough that could be broken, shared out and eaten as a collective ritual. My heart raced as I read how the cultivation and use of *huauhtli* was prohibited by the Spanish, so threatened were they by the indigenous reverence of this plant. Fields of the crop were razed to the ground as a cosmovision rooted in the earth was replaced with enforced Catholicism. I couldn't quite believe that this same plant, or rather its great-great-great-grandchildren, was now growing outside in our South England soil, revealing itself to us day by day.

The name amaranth, *amaranto* in Spanish, comes from the Greek *amarantos (αμάρανθος)*, meaning unfading, everlasting flower. It feels ever more poignant then that it is thanks to the unfading, everlasting love and work of indigenous communities in Mexico that the seed survives today.

What all my internet research back then couldn't tell me was that amaranth would be the catalyst for my returning to Mexico and reconnecting with my family there, or that this would be the beginning of a back-and-forth between these wildly different cultures that pulse through my veins. It couldn't tell me that five years later I would sit hunched over a keyboard in a small cafe in Oaxaca, where fields of amaranth grow proudly beyond the city, trying to put into words what has been, and continues to be, one of the most significant relationships of my life.

1.   Solnit, Rebecca. *A Field Guide to Getting Lost*. Edinburgh: Canongate, 2005.

2.   *Alebrijes* are brightly coloured Mexican folk art sculptures of fantastical creatures.

JACK GOODWIN

The Wriggling Mischief of a Dangling Bramble Vine

Somewhere between the out and
In-breath of the year
our bodies rest, energy pulled down to
the roots,
dormant, yet alive.

I am here intentionally; deliberately
slow, gazing over the
winter's edge. *I'll wait a little while here,*
let my porous skin
absorb this biting darkness.

Outside, the tang of fermenting apples wafts
off the cold ground.
Rock and Root bruise my naked feet and
piles of soft feathers tell me
Fox is nearby, nestled warm below.

Robin accompanies my slow
morning amble. Leaves
crinkle underfoot, mulching down.
How I wish I could lay somewhere
long enough to experience my slow
decay back into the ground.

Something grabs my beanie clean off my head
Expose yourself!
I hear, as I turn around incensed.
One thing I learnt on this Midwinter's morning:
Anger does not last long
in the face of the wriggling mischief
of a dangling bramble vine.

HEMPEN

Grow Hemp

Not for profit co-operative Hempen share the challenges of growing hemp

What plant can produce fibres strong enough to build houses, seeds rich in omega oils and flowers full of medicine; all the while sequestering more carbon per acre than trees? Hemp!

There is a long history of hemp cultivation in the UK that goes back to as early as 1533. For every 60 acres owned, a quarter was legally required to grow hemp due to its value for the nation. Hempen Organic Co-operative began as an idea to form a resilient rural community centred around the cultivation of hemp, as we believed in the versatility of this miracle crop.

This idea became reality when an enlightened tenant farmer on the Hardwick Estate agreed to plant an experimental hemp crop in 2014 with one of Hempen's founders. The very next year, 27 acres were sown and a bumper crop was grown. In 2015, Hempen was officially established as a not-for-profit workers co-operative for the equitable, ethical and ecological production of hemp products.

For two winters, a core group of dedicated Hempenites camped in a field at Path Hill Farm (Hempen HQ). We developed products

that we sold at farmers' markets, and attracted more and more volunteers, which has made Hempen what it is today. Our holistic business model ensures we work together whilst actively supporting the well-being of our employees, volunteers, community, partner farmers, and the land.

Hempen initially focused on the production of the valuable organic hemp seed oil, which contains all the essential amino acids the body needs but does not produce. It is still one of our leading products, and inspired a nutritious hemp food range. When we found out about the myriad health benefits of Cannabidiol (CBD) for a range of conditions and diseases, we began to harvest hemp flowers for the essential oils containing CBD, as well as seed and straw for our other products. It is a much more valuable crop and our range of incredibly popular CBD droppers massively boosted our business.

We then we had a huge setback. You need an Industrial Hemp Licence from the Home Office to grow hemp at scale in the UK, and

it is illegal to do so without one. Under that licence, we informed the Home Office that we were now harvesting CBD. But in 2018, the guidelines for growing changed, explicitly stating that hemp was not to be harvested for CBD without an additional special licence for THC (a psychoactive compound found in extremely low levels in hemp). That licence could typically cost £10 million or more, which made it totally unaffordable for us, and most farmers.

In 2019, midway through our growing season, with hemp growing as high as our heads; we were told that our industrial hemp licence would not be renewed because we had harvested flowers to make CBD in the past. We had told them what we were doing on all our official paperwork for three years prior, but despite numerous communications and visits in that time, they never said this was a problem. We found ourselves with multiple fields of unlicensed hemp and facing a lengthy court case, with up to 14 years in prison for those involved if we lost.

With pressure from the Drugs and Firearms Unit at the Home Office, which licences the crop, we took the painful decision to plough-in a perfectly healthy hemp field, with a retail value of around £200,000 in seed and straw products, and £2.4 million in CBD products. With our industrial hemp licence revoked, we've been reliant on our organic partner farms to grow the hemp for our seed and straw-based products ever since. It remains legal for us to sell CBD products and import CBD concentrates grown in Europe, but it is illegal for us to produce it in the UK.

The destruction of our crop drew international media attention and we used this momentum to launch a campaign to change UK hemp policy, which currently suppresses UK hemp production and the CBD industry as a whole. This is despite World Health Organization and United Nations legal guidance advising that industrial hemp in all forms should not be a controlled substance, and cultivation should not require a licence.

We are not the only ones affected by this. Despite hundreds of applications, there are only about 20 licenced industrial hemp growers, totalling just 2,000 hectares, meaning that businesses struggle to meet the demand for organic UK hemp products. In these times of economic, social, and environmental crises; it is a missed opportunity that this new green industry, that benefits UK farming and public health, is being stifled by government policy that favours the wealthy and well connected. It is obvious to everyone who grows hemp or sells hemp products that the current system is outdated and not fit for purpose.

Unnecessary red tape serves as a barrier to entry; for example, you can only apply for a licence between January and February, and you may not hear back until the end of April when it will be too late to sow most spring crops. If you have been rejected there is no appeal mechanism, and if you've been successful you still have to get seed from Europe which can take weeks. Farms get rejected for the most unlikely reasons — one farm had a renewal rejected because it was near an atomic weapons establishment that had been there when the original licence was granted. Another was rejected for being within a mile of a bed and breakfast.

Hempen now supports many farmers through the industrial hemp licensing process to grow a healthy crop of hemp that will benefit their soils, their finances, and the planet. We have learnt the hard way, so we can help new growers get it right the first time. Hempen and other hemp farmers are calling for this farcical licensing regime to be scrapped. At the very least, it should be moved from the Drugs and Firearms Unit at the Home Office to the Department for Environment Food and Rural Affairs (Defra), and treated like any other agricultural crop. Studies are begining to show that hemp is a great break crop alternative to oil seed rape, improving biodiversity and soil regeneration, yet another valuable use. If, like us, you believe in the power of this plant for the benefit of all, then get in touch. In these crazy times, what better plant is there to help build a healthier and happier world?

MARTIN GODFREY

Farming Through a Changing World

Organic grower Martin Godfrey from Hilltown Organics and Harvest Workers Co-op reflects on his land work journey

Growing up on a small mixed family farm in East Devon, the summers seemed long and consistent, unlike today's chaotic weather patterns. We grew and harvested swedes and spuds by hand. We picked apples from the old orchard trees with buckets and hooks, climbing between the bows of 30ft tall trees. At seven years old, I had a pet sow, and bred her to raise piglets to fatten for market.

I just about remember cutting 'flagpole' cabbages for the dairy cows with dad, they were bigger than me. The tradition of growing fodder cabbage has long since passed. The plants were brought in as brassica 'peg' plants then planted in prepared land; I even remember Granddad exhibiting them at the local ploughing match produce show. In the autumn, we would go to the field with a tractor and trailer for a fresh load of cabbage harvested by hand with a billhook for the 60-strong dairy herd. I was brought up learning the work ethic of a farmer's life rather than seeing work as a job.

One day, I was offered the opportunity to grow veg as a job, which had been my hobby since I was a boy. I moved to Mr Braggs' farm,

Shillingford Organics, on the edge of Exeter. This was my first
experience of producing food and cultivating soil the natural way,
without artificial inputs, and it was clear to see that the farm was
brimming with wildlife and biodiversity. From then on, I continued
learning about soil life and natural systems, as well as finding out
about food politics in the process.

I am 54 now and have worked through very different decades of
farming, when many changes were taking place. I have been right
through a variety of farming methods, but overall they were good
times for me. Today, I am a much wiser farmer. I clearly see the harm
we have done to our ecosystems and am very grateful to be farming
naturally, with the sense of repairing the early years of farming
conventionally.

For more than ten years now, my wife Sara and I have run our
smallholding; integrating organic, permaculture, polyculture,
silvopasture, wilding, and agroforestry on our challenging site on the
edge of Dartmoor. We have a diverse range of habitats with many
species of wildlife that were missing here before, and produce nutrient
dense food for our local community. Our growing system includes
wild edibles, which contain an important range of phytonutrients
otherwise missing from today's plates. We like to share the benefits of
foraged foods through workshops and our pop-up café.

As a younger man, I admit that I was blind to the evils of corporate
agriculture. Today, I am part of a co-operative working to rebuild a
better local food system by supplying local organic food, and raising
awareness about food and farming. Sara and I are both passionate
about nature and nutrition, and believe farming has become far too
industrialised, big, clinical, and dependent on corporate poisons.
That is why we got involved with the Landworkers' Alliance right
from the start.

Sara and I were staying at Jyoti Fernandes' Fivepenny Farm in Dorset
as volunteers, processing apples for juicing, when we, and others,
discussed the idea of starting a new organisation — an alternative
to the National Farmers Union. Sara became part of the core group
that originally set up the LWA, and its emergence as a proactive and

dynamic movement has been timely and crucial as a counterweight to corporate industrial farming tightening its grip on the global food system. It was exciting to be involved in the beginnings of the LWA and it's been incredibly satisfactory and uplifting to see it grow from strength to strength.

Root and Branch

BECKY DAVIES

Barddoniaeth y Goedwig

Poetry of the Woodland

Gwrandewch ar farddas y goedwig wyrddlas
Coesau cryfion pren
Yn crychu uwch fy mhen
A chanopïau dail
Yn siarad â siffrwd bob yn ail
Un coeden i'r llall
Mae'r adar yn deall
Ieithoedd deilen a gwraidd
A chyfiethant o'r gwraidd

—

Listen to the poetry of the verdant woodland
Strong limbs of wood
Creak above my head
And canopies of leaves chatter and whisper the one to the other
From one tree to the next
The birds understand what is said
And translate it into ancient song

JAY JORDAN AND ISABELLE FRÉMEAUX

No Commons without Commoners

Jay Jordan and Isabelle Frémeaux from
The Laboratory of Insurrectionary Imagination
provide an excerpt from their book, 'We are "Nature"
Defending Itself: Entangling Art, Activism and
Autonomous Zones'

Winter 2019. I feel like a clown slaloming between trees with my comically oversized orange boots and hard hat, and my fluorescent thick trousers. I follow my friends, awkward and happy: today is the first day of the week-long collective logging in the Rohanne forest and, equipped with the appropriate (albeit ridiculous) safety attire, I have joined the group of novices learning the art of taking care of this small but precious forest, to which we now belong. My dad, a lorry driver, would be surprised but proud to see his daughter confidently carrying a chainsaw on her shoulder. We are going to be under the guidance of a dozen experienced comrades who have come from collectives and forest-based struggles all over the country. As measured by the commercial timber industry, it is a patch hardly worthy of interest — 80 acres of 60-year-old deciduous and coniferous trees. But to us on the zad, it is a whole world of its own.

This is the forest where some of the fiercest battles took place during

Operation Caesar, with its tree house dwellers, mud slingers and blockading pensioners. It has provided timber to build some of the most gorgeous cabins and buildings on the zad, as well as firewood for the cold winter days. It has been the source of highly heated debates about the ethics of human intervention in 'nature' and the stage for theatrical candle-lit meanders by night. Left officially 'unmanaged' for years in anticipation of being eradicated to make way for the airport, it is today the subject of an arm-wrestle with authorities, adamant that they should regain full power over it. No agreement has been reached with the National Office for Forests, whose agents are the only ones authorised to extract wood in publicly owned forests. Walking in the footsteps of the commoners who came before us, whose survival was criminalised as 'poaching' by those who wanted to force them off this land, we are about to do what we do best: disobey.

'If you fell this chestnut tree, it will give more light to that young oak tree there, which is what you are aiming for. But you are going to have to be precise in your cut so as not to damage this other one on the way down.' We spend more time with our necks crooked, staring at the canopy and discussing with our friends-turned-trainers how to go about cutting which tree, than with our chainsaws in action. Our focus is put on taking care of the forest ecosystem, 'not as in protecting something fragile' (although it might be), Carmine reminded us at the start of the day, 'but in the sense of acknowledging mutual 'needs'. The aim is finding the right balance between our needs for timber and firewood and those of the forest, so that it can continue to flourish. Obviously, what constitutes this equilibrium is the topic of numerous passionate discussions among the members of the collective Abrakadabois (a playful portmanteau of abracadabra, the magic formula, and the French word for Wood, *bois*), who are dedicated to taking care of the zad's woodlands and hedges.

To move beyond these divisive conflicts, the group has committed to develop a shared vision and increased sensitivity through skill sharing, collective learning and a common appreciation of the forest. Since 2016, it has brought together passionate amateurs, an ex-forestry engineer, a gaggle of tree surgeons and lumberjacks; and has been organising reading groups to share knowledge and questions about plant biology and the latest research about mycorrhizal symbiosis and the communication between trees, as well as anthropological texts on inter-species collaborations developed by hunter-gatherer civilizations around the world. These conversations nourish the elaboration of a common (albeit manifold) perspective. This is enhanced by regular walks taken together to learn not only to recognize trees and identify possible diseases or specific behaviours, but also to analyse the impact that a previous cut has had on the growth and development of its neighbouring trees, how it affects the lives of insects, the paths of mammals, etc. This same group determines each year which trees will be cut. Through attention and observation, we thus learn the web of interdependencies that is life, and progressively sharpen our ways of seeing.

'A forest like this one is not interesting to the industry', explains Michel, who worked as a forest engineer before deserting seven years ago to live on the zad, 'It is too small, the trees grow too close together because they were left 'unmanaged' for years. If they had it their way and took back this forest's management, the most probable option for them would be a clear cut. Clear cutting has become increasingly common and violent: nowadays trees are not just cut down — stumps are dug out and the slash (debris) is taken away to turn into the supposedly 'ecological' heating source of biomass, even though leaving it in place would protect the soil and the aquifer and aid in restoration. Then a monoculture forest is planted: rows of fast growing trees on impoverished soil needing fertilisers (copper, phosphorus) that end up in drinking water!

Against this extractivist logic, Abrakadabois has been learning from and networking with folk throughout France; along with researching, practising, and defending a silviculture that does justice to the inherent dynamics of the forest. As philosopher Baptiste Morizot describes, by taking the point of view of the forest, their practices are 'full of regards for it'.[1] This soft silviculture aims to work *with* the forest rather than against it, caring for its limits and ecosystems, extracting wood while preserving, even restoring, the soil and the tree health, and respecting the microhabitats with a careful holistic approach that recognizes that we are not in the forest; we are part of it, and it is part of us. It is all about progressively forging an 'alliance of needs' between humans and more-than-humans, made possible by the diversity of approaches and ways of seeing: naturalists and lumberjacks, amateur tree lovers, and professional foresters, sawyers and inhabitants.

1.   Morizot, Baptiste. *Manières d'être vivant*. Arles: Actes Sud, 2020.

TOM KEMP

Community of Foresters

Forester Tom Kemp from Working Woodlands discusses restoring our woodlands and being part of the LWA's network for foresters

The west of Cornwall is pretty isolated geographically and forestry is relatively underdeveloped; it is easy to feel removed from the bigger picture here in this ancient Cornish wood.

Situated between Truro and Falmouth and dating back to at least the 1650s, Devichoys Wood is oak coppice with hazel, holly, and rowan; plus some rarer alder buckthorn and wild crab. Mosses and lichens hang off many of the trunks, and old charcoal platforms notched into the hillside tell of the previous foresters who made charcoal for the tin industry locally. It is owned by the Wildlife Trust, who tried for some years to reinstate coppicing themselves. Regenerative forestry requires a lot of specialist skills, equipment, and markets though, and they ran out of steam to continue this work. When our small agroforestry business approached them looking for a woodland to restore, while generating an income from the produce, they were really keen. We now have a good working example of a conservation body and forest contractor relationship that delivers high quality nature recovery and is self-financing. It is a true privilege to bring this wood back to life. As we restore the coppice, we have seen a big

increase in birds, flowers, and butterflies; and bats when we check the charcoal kiln on summer nights.

Our aim is to bring Cornish woods back into active management, to provide products and services for a post-oil future. We work with a wide range of regenerative forestry methods, from traditional techniques such as coppicing, to more recently developed methods such as continuous cover and agroforestry. We want to make our woods places that people engage with in a deeper way than dog walks, a dynamic, thriving ecosystem that locals are proud to know. A place that provides their firewood, building timber, and more. A place where they can talk with foresters and understand how important their local natural resources are to their daily lives. Devichoys Wood is open to the public, which means we close sections while felling and keep a keen eye out for wanderers. It also means we get to meet local walkers and talk about forestry — and sometimes add them to our customer base! Overall the often negative perception of tree felling turns to discovery and education when people get to chat with the foresters. They find out about the many amazing uses trees provide; and come to understand the benefits of diversifying a wood's canopy for biodiversity, carbon, and climate resilience.

Here at Working Woodlands Cornwall, Nick, Grace, Emma and myself feel the need to reach out and be part of a broader movement towards connecting with woods, trees, and the places that provide timber to hold up our roofs. One of the ways we do this is through our podcast on regenerative forestry, *Future Woods*, which we started during the pandemic. It is a way to maintain a connection for people interested in regenerative forestry, and give a space for the varied approaches to managing woodland well. Another key way is being an active member of the Landworkers' Alliance.

Gathering with other LWA members in May 2022 at Hillyfield in Devon for the first national gathering of the forestry branch, and hearing from other woodland workers who are milling their logs into homes, producing firewood for their local communities, and laying hedges to keep in their livestock; I felt that there are many of us working to create the essentials for a happy life. Through talks

and over long table meals and evenings of music, dancing, and full glasses; we had a strong sense that we belong to a really good community of woodlanders. As we shared the challenges we face and the wider picture of our work in relation to changing trends in land management across the UK, we felt part of a movement for change.

The three hours in the car back to Cornwall were filled with enthusiastic talk about the weekend and our renewed drive to create greater impact through our work. Being part of a connected community of regenerative foresters, agroforesters and farmers makes a big difference to the work we do and why we keep doing it. Working in the woods and running a business requires real physical and mental effort, often long hours, getting home cold and wet. Being part of a growing community and movement is a reminder that I'm not the only one doing this; hard work is part of the process of making the world a better place to live, and thousands are here alongside me. Knowing there are others to bounce thoughts off and share experiences is exciting and feeds my inspiration. Knowing who to ring to chat over specific ideas helps my development as a forester, but also leads to proper warm friendships and makes for a more meaningful life, which is what most of us do this work for when it comes down to it.

Growing and Planting Trees in South Devon

FLAME member Sky Miller shares his experience of working with trees

2021 was the start of my landwork journey. It was a year of transition, when I finished college at Exeter — where I studied engineering; and started my first job as a trainee with South Devon-based native woodland restoration charity, Moor Trees.

Throughout the year, myself and nine other trainees learnt the whole process of growing trees, from seed collecting from parent trees right through to planting out at many different places across South Devon (including some on the Dartmoor Commons). I came to really appreciate trees; and the more I worked with them, the more mind-blowing I found them to be. What was particularly fascinating to me is how most grow from seeds that are often just a few millimetres in diameter, into extraordinary giants (some good examples include oak, beech, and yew). Some trees, if left to their own devices, will live for many hundreds of years, and on rare occasions over a thousand!

We collected seeds from across different places in South Devon; often woods and hedgerows on public footpaths or public access land, and sometimes from areas where permission had been granted for us to collect. All seed sources had to be documented to ensure any diseases could be tracked; Moor Trees would often go back to places we had collected from before, as we knew the trees were in a good state. Next, and the most time consuming job, was processing the seeds we had collected. For some seeds, this entails removing the juicy outer layer so the seed itself doesn't end up rotting, which would mean no trees. Removing this layer is not anywhere near as easy as it sounds! If done properly though, the results are well worth the effort. It is worth noting that animals have perfected the task already; they consume the seeds and then excrete them without the juicy layer (but who wants to sift through bird poo all day?!).

The seeds needed a suitable soil environment; so we created a mixture of peat-free compost and sand, then layered the seeds into plastic bags which would be checked regularly for moisture content. Between processing and germination, we mimicked a cold winter for some of the tree seeds so they would germinate, by putting them in a fridge. Once they were showing signs of germination, it was time to very carefully transfer them to seed trays, before transferring to larger containers to grow on to a suitable size ready for planting. This growing stage at the tree nursery usually takes about two years, during which the seedlings need plenty of care and attention; including watering during the summer, checking for various leaf diseases like rusts, keeping rodents out, etc.

Whilst at Moor Trees I was able to go on two placements to local projects: The Living Projects at Pondfield, a local land based community space with a small market garden, where I helped with all the various tasks throughout the growing season; and a woodland farm called Hillyfield, where I learned many different skills including coppicing, fencing, and getting my chainsaw certificate.

It was during my traineeship at Moor Trees that I joined FLAME (Food, Land Agriculture: a Movement for Equality), the youth branch of the Landworkers' Alliance. Being part of FLAME has helped me to meet a great bunch of like-minded people of similar ages to me (especially at the Land Skills Fair). I like that everyone involved in FLAME is passionate about the land, and how we can make it better for nature and for people.

As well as connecting with other people, FLAME has lots of opportunities to gain new experiences; one that I got involved in was a hedge laying course with The Heart of England Hedge Laying Group. The course taught practical skills in how to lay a hedge in the Midland style (across the country there are more than 30 different styles, almost every county used to have its own hedge laying style). The Midland style involves pleaching the stems to 45 degrees and laying them between stakes, the hedge is finished with a woven hazel binder at the top.

Since joining FLAME, and the LWA as a whole, I have been involved in meetings, including a recent South West regional assembly; and I have also been introduced to other interesting organisations, such as Farming and Food Education (FarmED) — in particular the developing Emergent Generation — and Groundswell The Regenerative Agriculture Festival.

I finished working at Moor Trees in April 2022 when the funding for my role ended; but the experience of working with trees, learning how to nurture them, and understanding their remarkable qualities, means that I'll always have an avid interest. Looking to the future I hope to be able to work in the areas of conservation and regenerative farming.

Although I have switched very early away from engineering, I feel it was a good decision. I have always felt a close connection to nature, and am looking forward to using the skills I gained whilst studying engineering in my conservation and regenerative farming work.

LUCY GROVE

Coppice

The first cut, always by hand.

Metal and muscle meeting time through growth rings —
Remember that soaring sun the spring before last?
That wild, wet winter?
That bumper beech mast?

Muscle and metal, we are tracking back time now,
Back before we were here.
This coppice has been forgotten many a year.

Years have passed since the last dormouse dozed through winters
deep, tucked up asleep.
Decades gone, a silenced song, from the last coppice worker taking
billhook to bark, making his mark.
Memories have faded of the last heavy horse pack, heaving timber
down the old rutted track.

But — with whiskets of biscuits and sloe flavoured rum,
We are returning again, the land-loving ones —
See with axe and with saw,
We are reclaiming our lore.

We watch life spring from each old stool,
Our coups creating mosaics of fuel.
Poles are pitched to waiting hands,
Hurdles crafted for neighbouring lands.

See this is our regeneration, our re-imagination, our stand —

Of a life longed and lived through graft and laughs,
Of light lit leaves and charcoaled sleeves.

The last cut, always by hand.

PAUL COOKSON

Our Coppicing Journey

Green woodworker Paul Cookson shares Green Aspirations' efforts to bring together communities to revive coppicing in Scotland

Coppicing is one of the earliest forms of woodland management and has provided a sustainable source of timber for thousands of years. Coppice is created by cutting trees to ground level, creating a 'stool' and encouraging new shoots to grow. When coppice is actively managed, it produces straight sticks that we have found hundreds of uses for, including fencing, hedging, baskets, and building. Coppicing was a major industry until the First World War, when industry replaced labour intensive natural materials with cheaper synthetics. This led to a huge decline in coppice being cut, and the Forestry Commission described it as functionally dead in 1926. Since then, there has been virtually no coppicing — so what do you do if regular customers ask you to provide hazel hurdles, but there is no local hazel to buy?

Around eight years ago, we were asked if we could either supply hazel hurdles or teach people how to make them. We looked to see what coppiced hazel was available here in Scotland, but with no luck; the limited coppice that existed was being cut by individuals for their own use. With no success in finding coppiced hazel to buy, we decided that it would be worth trying to find some coppice to restore, or bring back into rotation. We put the call out on social media and soon started working with South Lanarkshire Council.

The site is a park that was largely planted up around the millennium, when large swathes of the Scottish central belt were planted. For some reason, the forester that drew up the plans added a 1.75-acre area of hazel. The recommended minimum annual cut is 0.25 acres, so this gave a seven-year cutting rotation. There was a map showing these rotation areas but not much else; and the hazel had not been cut for its first 14 years. We had to learn fast and settle in for a long haul.

Coppicing takes place in winter, so that the trees are dormant, and the cutting doesn't impact nesting birds. Cutting old hazel by hand and carrying it all out along a muddy track is not for everyone — this is physical work in the depths of the worst weather. But to some people, this seems the most natural way to spend time in nature, after all we have done it for thousands of years! Despite all the hard work — much of it a blur — we are now seeing the benefits. Initially, all the overgrown hazel was used for charcoal; but, as I write this, we are recutting that first year and getting some great hazel sticks to use as products like hurdles and bean poles. All the old growth has been cut and the site is now back in rotation, on its way to full production — probably in the next rotation or two.

So, what have we learned? Each year, we have improved how we work; learned how to process the timber, and to build markets for coppiced sticks. We know that it could not have been achieved without lots of people dedicated to coppice restoration. In the first year, we worked alongside a local scout group. Since then, we have built up a regular — and very hardy — group of volunteers to do the work. It has taken well over 100 days each winter for seven years to get this far. The 'cutting crew' is now a regular volunteer group, and has provided a way to spend time together outdoors through the tough past couple of years.

Restoring coppice is hard work, and needs a long-term plan to keep in rotation if we want to have good quality materials that are sustainable. But it is well worth the effort. Not only have we seen an area of woodland transformed, we have shared traditional skills; seen biodiversity increase, with new plants and butterflies; and have a sustainable, continuous source of a versatile material. And, as we have recut the first area this year, we have halved the number of days since the initial cut. So, now it's on to a new restoration project.

Hand in Hoof,
Hoof in Hand

UNKNOWN

Three Acres and a Cow

You've heard a lot of talk about three acres and a cow
And if they mean to give it us why don't they give it now?
For if I do not get it I may go out of my mind
There's nothing but the land and cow will keep me satisfied

Don't you wish you had it now, three acres and a cow!
Oh you can make good cheese and butter when you get the cow.

There's a certain class in England that is holding fortune great
Yet they give us all a starving wage to work on their estate
The land's been stolen from the poor and those that hold it now
They do not want to give us all three acres and a cow

D'y' think they'll ever want to give three acres and a cow
When they can get a man to take low wage to drive the plough
To live a man he has to work from daylight until dark
So the lord can have both bulls and cattle grazing in his park

But now there is a pretty go in all the country though
The workers they want to know what the government will do
And what we have been looking for, I wish they'd give us now
We're sure to live if they only give three acres and a cow

If all the land in England was divided up quite fair
There would be work for everyone to earn an honest share

Well some have thousand acre farms which they have got somehow
but I'll be satisfied to get three acres and a cow

Lyrics in the public domain from a broadside from the late 1880s
and featured in *The Three Acres and a Cow* songbook.

NIKKI YOXALL

Hand in Hoof

Nikki Yoxall from Grampian Graziers writes about using agroecological principles to produce beef from rare and native breeds in North East Scoland

The world we find ourselves in can feel chaotic and scary. We are faced daily with the challenges of climate change, health pandemics, conflict, and biodiversity loss on a massive scale. It's easy to become caught up in the fear, anxiety, and overwhelm of these issues; these threats.

To help me manage my feelings of despondency around a perception of global inaction, I turn most often to my cattle. These slow, calm-eyed beasts who I share my life with, who depend on me for access to feed and water, and who I depend on to heal and restore the land in this corner of Scotland we call home. This reciprocal relationship, where I make sure their needs are met and they in turn act as ecological engineers to shape and mend intensively managed pasture, slows our lives to focus on the simplicity of these actions.

Running a grazier business with my husband, we graze our livestock on land owned by other people, helping them to reach their ecological goals whilst producing 100% pasture and tree-fed beef. We describe our farming approach as agroecological, this means we aim to optimise the interactions between plants and animals, integrating our herd of cattle into the ecosystems we have the privilege of becoming a part of as we graze them. For us, farming isn't something that happens in a defined space, with nature taking precedence elsewhere. Instead, our cattle nudge and enable ecological processes to happen, they become part of 'nature'. We aim to think with and for the land, considering not only now, but the future for the soil beneath our feet and the plants that grow here. And that consideration for land goes hand-in-hoof with managing and caring for livestock.

Our herd is made up of a group of characters, all of whom have names, but most of whom will one day enter the food chain. We care deeply for these animals — Georgia the herd matriarch who always keeps a watchful eye on the calves, Betty the lead cow who scopes out the perimeter of new paddocks and always finds water, Alder with her scraggly rigget hair, and Badger, the smallest calf, who always brings up the rear when we move the herd; to name just a few.[1] For some, this juxtaposition of love and respect for an animal and the inevitability of death and subsequent eating is difficult to understand. For us, it makes sense to fill every moment of their lives with care and compassion, whilst having the deepest respect for the nourishment they give us. When we look to the ways of ancient hunters, who gave thanks at the kill and were bound in reciprocity with their ecosystem, we feel a deep connection and understanding that it is possible to be nourished, both emotionally and physically, by these animals. This act binds us with the ecosystem. Our food is of this place; it drank the same water as us, ate plants from the same soil in which our vegetables grow, we nourished them in life, and they nourish us in death — the cycle continues as all cycles must, we are just part of the ecosystem.

In fast-paced times, where social media reigns and speed is king, where immediacy feeds our insatiable hunger for stimulation; taking a breath and thinking about our connection with our animals, our home, and our food is a soothing balm, and key to developing resilience for an uncertain future.

1.   'Riggit' Galloway cattle are a strain of Galloway, identifiable by the white stripe running down their spine. The term 'riggit' is a Scottish vernacular reference to this stripe.

GEORGIE STYLES, ISOBEL TALKS, AND ADAM SCARTH

New Entrant Interview

Adam Scarth, new entrant farmer on the Kickstart Scheme, interviewed by Georgie Styles and Isobel Talks

GS **Could you introduce yourself, Adam?**

I am Adam Scarth; I'm 20, from Leeds, working at Swillington Organic Farm. I got into that work through the Jobcentre, a Kickstart programme that I wanted to do because I wanted to work with animals. I had already done two years on a collage course in York, so I had a background in working with animals. Plus, I've had work experience in animal places, like horse centres.

When I was younger, my ultimate dream was to be a vet. The qualifications for University are holding me back at the moment, so I'm just trying to get as much experience in animal care as possible — working with small and big animals.

GS **Do you want to work specifically with farm animals?**

It is looking that way. It seems to be a lot more fun with them, more of a task, which I like — I like challenges.

GS **So tell me about the farm you're working on.**

We have sheep, pigs, cows, horses, ducks, geese; it's a big farm. There are fishing lakes there as well. It's in Swillington, just outside of Leeds.

GS **Can you tell us about the Kickstart setup?**

It started with a six month contract with the farm, working in the garden and planting, then digging stuff up for a stall, which is run by volunteers on Saturdays. It's a veg stall, and people can also come and collect eggs.

The usual morning routine is to water all the stuff in the polytunnels and greenhouses. Then I walk down to the chickens, collect the eggs, and let them out because they're free range. Then see what's next, it could be looking after the game chicks, feeding and watering them; planting stuff in the garden; or a day with animals.

GS **Do you enjoy both working with animals and growing?**

Yes, but I would rather be with animals all day, every day. I can't be picky though, as a farmer you need to manage your crops as well.

GS **How often are you at the farm?**

Four days a week, Wednesday to Saturday. I've been working there for just over a month, with another worker called Ryan, who's been there for around six years now. There are also two butchers who have their own set up. The pigs and sheep go to the slaughterhouse, then are brought back so that all the butchery is done on the farm. The chickens are slaughtered and butchered on site. There's a shop on the farm too, but as soon as the animal's killed it's nothing to do with me!

GS **Are your family involved in farming?**

No. I'm originally from Leeds, a 15-minute bus ride from the city centre. It's like an estate — houses, a shopping centre, stuff like that. No farm land; the closest farm is about a half hour walk away. I went there as a kid and had a look around.

GS **Can you tell us how you came to work on the farm?**

After school I went to agricultural college to do animal care. I did two years there, but it was just a bit repetitive so I didn't see the point in staying on to do the third year. I did my GCSEs again, and then I got Level 1 and Level 2 Certificates for Animal Care. I learnt how to do health checks on animals, from head to tail; starting with domestic animals, then we went to farms. We had cows and sheep at college so we worked with them too. You can go into any type of work — become a vet, work in a pet shop. Someone on the course went on to work at the Dogs Trust. I did some voluntary work at a pet shop, then to earn money I did warehouse work for two years. I got bored though — it wasn't for me, so I went back to the Jobcentre.

At the Jobcentre they asked what I wanted to look for, I said I wanted to be a vet. They had just had a job come through, and I applied straight away. I got an interview three days later. After that, I was told that it was a close call between me and another person. I thought that's it — I wont get it, this person will have one up on me. But because of the two years of practical work at college, I got the job. I was happy! I hope I can stay on. When it comes to my last month, I'll ask if I can stay on a further contract.

GS **What do you want to do next?**

I'd like to stay at the farm. A vet was called in on my last shift because one of the cows calved and her insides came out; so we had to push them back in, then they came back out, and the vet got me involved, doing the injections and everything.

It was really fun! It was the first time that I had done that, and I could ask questions — 'how do I know if I got it in the wrong spot?' I learnt a lot. If you do it wrong you could seriously harm the animal or kill it. I'm going to talk to the vets that come out, see what would be best for me to do next, then I can move on to where I need to be. I like being on a farm, you're not stuck in one place, you're out in the fresh air. And especially organic farms — they don't smell like normal farms because of the food resources they have, nothing artificial.

GS **Did you know anything about organic farming before?**

No, I'm learning everything at the farm. It's a job that I'm getting experience and education from, all in one place.

IT **Is there anything else in organics that interests you?**

Yes, it's because they're not fed food daily. They have organic pellets, but only if pigs are having piglets, sheep are lambing or the cows are calving — they only get feed when they can't be out grazing. The price of organic meat is higher than it would be from a normal farmer. It's weird, because it's all still the same meat, but the way it's raised means there are more nutrients, which puts the price up. I'll eat any meat though. The butchers made sausage with wild garlic and white wine — it was really nice. It was supposed to be £4.50, but I got it for free — perk of the job!

IT **Did you face any challenges to get on the path you're on?**

My GCSEs knocked me back. At school I didn't knuckle down, so I had to retake them all at college while doing the animal care Levels 1 and 2. But as soon as I got through that work, it goes as it flows.

IT **Was it frustrating to do warehouse work?**

Yes, it was boring and I didn't want to get up for work. But now at this farm, I am up! The other day, I knew we were calving and usually I wake up at 7am, but I woke up at 6am and I couldn't get back to sleep — I was so excited! I didn't think calves could be so strong, I got pulled over by one. With the sheep, we had lambs come, then the mother died, so we bottle fed them. When they see you now, they now run over to you — it's amazing.

GS   **What is your motivation?**

My motivation is working with animals. I love it. Making sure they are happy and healthy, watching them grow. It's like watching a baby grow, I raised my nephews.

IT   **What do your family think about this type of work?**

They think it's amazing, it's something different and that's what I like about it. Everyone else in my family has done warehouse work, my dad is a painter and decorator for Leeds City Council. I'm going absolutely my own way. They love the fact that I'm doing something different, I've not followed in anyone else's footsteps. I've taken my own track, set my own speed. My nephew looks up to me, maybe he'll take my path. One day maybe I'll own my own farm; I'd love to own my own farm. I wouldn't need to pay for the vets because I'd be one myself. It would be out in the middle of nowhere, with one part that's away from the animals with a motorbike track to rent out. I'd have crops near there. Then on the other side, there would be animals and farm houses, that would be amazing. It would probably be in Leeds, around the Leeds area, that's where my family is.

GS   **In terms of your vision, to be an AgVet, do you foresee any challenges in getting there?**

Yes, getting into University. I need to find the right University, then be accepted. I'm not the brainiest of people, but I can take things in and learn quite quickly. The other day I learnt how to milk a cow. She's got a calf, but she has one teat that's bigger than the other and the calf can't get its mouth around it, so every day we have to get the milk out. You can be there for 20 minutes and it's fine, then your wrist starts aching and aching. You need to do it daily otherwise the milk will go off in the teat and come out lumpy.

GS **Would you drink the milk straight into your tea?**

Yes, I'd try it! It is full of nutrients. Like the eggs, on a daily basis, straight out of there and straight into a pan — I get to bring them home every week. I carry in 300 eggs in one day!

GS **What are the requirements to get on to the courses?**

Maths, English, and Biology. I did Maths and English at college. I haven't done Biology; there was some in animal care, but I didn't get the specific level for that. I would have to apply for grants and loans. I would go to Hull; get student accommodation, meet new people — I love meeting new people, telling my story, they can tell their story, and it's just amazing.

GS **Are there any other challenges?**

Finding land if I wanted to own my own farm; and getting a job that would pay enough to save and invest into that. I'd rather be a vet than have my own farm though to be honest. I would need my driving licence, vet licence. If I worked for a vet company I'd get my licence through them.

IT **If you could tell the government what you need, what would you ask for?**

A grant to get into University. The Kickstart is like an apprenticeship, but without college.

GS **If the Kickstart was longer, and it was available, would you want to do it — would you be interested?**

Definitely. I wish it was longer because I would learn a lot more. In six months you can learn a bit, but it would be upsetting to then have to go. I'd like to stay and learn more.

KATIE ALLEN

Farming Fashion

Farmer and designer Katie Allen writes about her knitwear collection made using the fleeces from her flock, and the need for fashion to connect with nature and the land

It is the first week in July and I am up early to gather the flock. It is a day with mixed emotions; the shearer is coming. The fibre we remove from the sheep today will be the start of my next collection of knitwear. I will be turning their bulky fleeces into beautiful handcrafted woollens, to bring warmth and a sense of contentment to people.

To be there for every stage of a garment's journey is undeniably fulfilling and it puts an extraordinary depth into my work. I love farming and it's a life I could not give up, however I always get apprehensive about a big gather. I never know if the flock will work into the pen as planned and pushing them through the race will be tiring. I will come away battered, bruised and exhausted.

We start the gather early, so it is relatively cool; I relish the opportunity to be pacing the fields, the tall pasture whipping at my legs. Our farm is located in North Wiltshire and as we chase through the flat fields, our steps flushing insects and birds up from the long grass, I can see the Marlborough Downs soaring up ahead of me. I know that wool has amazing qualities, which make it insulating when cold and breathable when hot, however I can't help but look at their enormous fleeces and think they must be sweltering under there — especially when they refuse to enter the pen, and instead race off in the opposite direction, their lambs hurtling behind, keen to keep up.

Solar energy has been utilised by these clever animals to grow a thick, fine fleece; and carbon from the atmosphere has been drawn down through the grasslands and locked into their fibres as part of their grazing and energy conversion process. My flock is made up of about 180 native breed sheep — two rare breeds: Castlemilk Moorit and Portland. As I watch them move through the fields, calling to each other and encouraging their young forward, it feels like the most natural fibre I can imagine. The domestication of sheep over the last 10,000 years means they need our help to remove their fleece each year. It's always a big relief for me to get the wool off their backs, removing it drastically reduces the risk of flystrike. This year I didn't use any preventative treatments as I want my fleeces to be certified organic and it put a lot of pressure on our management system. I think we've been lucky on low fly numbers though and our efforts have paid off. I will be pleased to see the wool sacks bursting at the end of the day and the flock bouncing around the field, lighter on their feet.

Like me, many farmers are keen to get the wool off, but they often don't share the same level of anticipation. When I bought my starter flock, I knew that their wool needed to become a valuable part of my farm business; it has taken more than five years of training to learn the design and craft skills that now enable me to create my knitwear. Not all wool sacks are full of the same potential. For me, shearing day is the beginning of the field to fashion journey but for most it is just a back breaking, expensive day resulting in a lot of worthless fibre. The farmers prepared to invest further time and money into rolling, packing and delivering their fleeces to the local depot are unlikely to cover their costs — the sale of their fibre will just help to offset the financial burden of the wool clip. You can understand why farmers, who already know they are making a loss after paying for the shearer, find it hard to justify the additional cost to prepare the fleeces and transport them to the depot. So instead they cut their losses and burn or bury them. I would like to say that I'm hearing less of these sad stories, but I'm not. Farmers are not any more optimistic about the value of their wool clip than they were five years ago.

It is different for our farm business though. As I look across the flock, laden with their beautiful fleeces, I can see right in front of me what an important part of the ewe's work growing her wool is, and imagine how much energy she must use to produce it. It should not just be a bi-product of our hogget and mutton. My flock produces more wool each year than lambs for market, and even my retired ewes can give me valuable fibre.

But I know that my journey is not possible for many farmers — I think that is why connecting farmers, processors and designers is so important. Projects like Fibreshed, that develop regional fibre systems, enable an honest mechanism to establish supply chains that really see and understand the people and processes involved in creating our clothes.

By the end of the day, my muscles ache, I am covered in scrapes, my nails are thick with dirt, and my hands greasy with lanolin. But the wool sacks are full — far too heavy for me to lift on my own. I look out from the shed and see the ewes with their heads down in the long pasture, their lambs playfully charging around in gangs. The sun is shining, the hedgerows are bursting with energy, and the wildlife on the farm is singing. Fashion businesses urgently need to reconnect with land and nature and our farmers and growers are the key.

KATE SCOTT

The Shepherdess

Shepherdess Kate Scott offers a short reflection on shepherding

She felt at one, enraptured by her surroundings as she sat a while looking out across the hills, where the mist clung to the soil like a duck down quilt with paisley swirls, and still hung in the nooks and crevices of the landscape beyond. Her mind cleared as the sun broke through, as if the morning fog had permeated every part of her being, dulling her thoughts and weighing down her limbs, seeping into the very core of her being and making even the birdsong on the hill sound January grey.

Grounding herself with her hands on the grass, she drew in the crisp sharp winter air as she watched the sheep in her care methodically munching the dew-laden pasture, looking up occasionally to question her presence. She could see the gentle curve of their bellies growing by the day. Not long now before new life, new growth, her favourite time of the year would soon be here, strengthening her inherent love of the land. A heritage of drovers, herbalists, and horse dealers had subconsciously shaped every aspect of her being, their lives being inextricably woven into the rich tapestry of her own.

SAMSON HART

Tzalmavet / Atah Emedi

THE SHADOW OF DEATH / YOU ARE AT MY SIDE

In darkened winter moments,
when the earth is cold,
and the days are short,
I think of my ancestors.

The ones who carved through the ice,
and built a fire for me to sit by.
Who carved through time,
a life for me to live by.

I see feathers plucked in the garden,
broth on the flame,
bread in the oven,
tradition in the making.

There is joy and there is pain,
they knew this more than me.
Now I also know it by the
songs they sing through me.

*Though I walk through
the valley of the shadow of death
I fear no harm
for you are at my side.*

Winter sun now on my face,
Spring, life, on the horizon,
there will be harvest on the table,
and you at my side.

HOLLY GAME

With the Bees

*Holly Game on learning from bees at Spiritwood
in Pembrokeshire*

I am fairly new to the world of beekeeping; and feel proud to be part
of a chain of women that has passed down the traditional skills and
knowledge of this rich craft, from generation to generation.

During my time at Spiritwood, I have helped tend to around 50
colonies for two seasons, each with its idiosyncrasies. I have been
fortunate to experience a myriad of weird and wonderful things that
can go on inside a hive. Despite this the bees are still teaching me
something new on almost every visit, constantly teaching me to be
humble, open-minded, and patient. As much as the ecologist in me
would love to leave them to their own devices, practicality requires
regular hive checks throughout the warm spring and summer months.

No matter what else is shouting for attention in our lives, or what
topics we have animatedly been discussing on our walk over to the
apiary, once we are suited up, we take a breath and ground ourselves
in a sense of calm and quiet. We often tell ourselves that this is for
the sake of the bees — to minimise the disturbance to them, but it
is equally valuable to us. Moving from hive to hive is a meditative
process of gentle repetition. Drawing your focus in allows so much

of the internal chatter to fall away. On a personal level, as someone exasperated by recent events, who struggles with anxiety and agoraphobia, beekeeping provides a safe and healing space that I can step into. My mind quiets to the task at hand.

The bees continue their important work following the familiar rhythms and cycles of the seasons, oblivious to a convenient human schedule. As I'm sure is familiar to many in landwork, they do not stop, so we cannot stop. For me, that stability and rhythm has been a lifesaver, a valuable reminder to breathe and work at their pace, rather than become preoccupied with what suits us.

Since working with the bees, my attention has become more in tune with the changing seasons, viewing the plants around me as potential sources of forage and observing the blooming cycles and floral senescence. Whereas before I would simply notice signs of transitions between seasons — the first buds of spring, the changing colours of autumn, or the arrival of a certain much-loved plant species; this has become a daily task of observation and recording. I have been inspired to spend time researching which flowers and floral features are beneficial to which bee and pollinator species, which are valuable for pollen or nectar, and which will have likely lost pollen in stormy weather. Using this knowledge, I then attempt to interpret the pollen colours and varieties safely stored within the comb. Endeavouring to view the world through the bee's perspective has only proven that there is so much more to learn. It drives my curiosity to discover more.

Solidarity
in Action

BECKY DAVIES

Hau Hadau

Hau hadau mewn gobaith,
mewn gobaith a ffydd
mewn gobaith a ffydd
Pan ddaw amser cynhaeaf ein 'sgubor llawn fydd
Hau hadau mewn gobaith,
mewn gobaith a ffydd

Hau hadau mewn gobaith,
mewn gobaith a ffydd
mewn gobaith a ffydd
Pan ddaw'r amser i falu grawn ein pobtai'n boeth fydd
Hau hadau mewn gobaith,
mewn gobaith a ffydd

Hau hadau mewn gobaith,
mewn gobaith a ffydd
mewn gobaith a ffydd
Pan ddaw amser rhannu bwyd ein calonnau llawn fydd
Hau hadau mewn gobaith,
mewn gobaith a ffydd

Refrain: Sow seeds in hope, in hope and faith

Verse 1
That when harvest time comes our barns will be full

Verse 2
That come the time to mill the grain, our ovens will be hot and ready

Verse 3
That when the time comes to share out our food, our hearts may be full

—

The idea is that the song leader sings the words that change in each verse, and everyone sings the repeating line whenever it pops up. Thus the song becomes a 'call and response'.

The mention of faith (ffydd) is to be interpreted differently by the heart of every singer. It could mean faith in the human endeavour of farming and sowing seed; equally it could mean faith in the earth and in the fertility of the land. Or it could tap into someone's faith in a Creator.

There is no one way of singing this song that is more or less valid than any other.

—

Song lyrics by Becky Davies for the Wales (Cymru) branch of the LWA, known as Gweithwyr y Tir (GyT) in Welsh.

DAN ILES

From Krems to Hebden Bridge: Landworkers' Alliance's Global Roots

Dan Iles provides an international context for the emergence of the Landworkers' Alliance

The Landworkers' Alliance was formed at the crossroads between the global and the local. A series of international gatherings brought together ideas and people, which ingrained the globalised perspective of the organisation we see now.

In 2011, I travelled to the small Austrian town of Krems an der Donau to take part in Europe's first food sovereignty gathering, which was inpired by the world's first such meeting in Mali four years earlier. There were strict quota rules to make sure each country was sending the right mix of grassroots activists, NGOs, land workers, and community representatives. A scattered call out for delegates from the UK ended as a collection of friends returning home with the aim of bringing about a national movement for food sovereignty. A movement which helped precipitate the emergence of the LWA.

Food sovereignty is a global concept with a global history. The phrase suits its Spanish and French translations a lot better; the meaning is more akin to autonomy and independence and less tied up with nationalist connotations.

The principle of food sovereignty, a concept which at its core tries to redistribute the power concentrated in the food system, was first launched by the global peasants' movement La Via Campesina (LVC) in 1996 in Rome, during the UN's Food and Agriculture Organisation's World Food Summit.

At that meeting, non-governmental organisations saw the power of fostering transnational connections between peasant and community movements. They wanted to link LVC with Indigenous peoples, women's groups, trade unions, and youth and environmental organisations from around the world.

This came to fruition at the world's first international food sovereignty meeting in Mali in 2007, where 500 people from 80 countries established the Declaration of Nyéléni. The name 'Nyéléni' was chosen as a tribute to a legendary Malian peasant woman who farmed and fed her people well, and the declaration agreed to some of the key food sovereignty principles many of us are still fighting for. The gathering also dreamed up the need for regional gatherings across the world.

One of those regional meetings took place in Krems an der Donau. 400 delegates from over 40 countries attempted to globalise a plethora of local movements that wanted to take back control of their food systems. The conference was a hive of skill sharing, policy debates and dreaming, which tried to bridge Europe's many divides — from East to West, rural to urban, anarchist to social democrat, young to old.

It was an occasion to laugh, connect with kindred spirits from different walks of life, drink wine and beer at a sliding scale based on the purchasing power of the country's currency, and most importantly, celebrate Europe's food culture.

The overarching aim and agenda was to agree on the wording for a European statement and action plan on food sovereignty, but, as is so often the case at these events, the main outcomes were measured by the weight of social fabric woven between the delegates.

I attended on behalf of Global Justice Now and remember evenings full of conversation, long debates, dancing, and sharing delicacies. It was at this conference that many of the UK delegation met each other for the first time, and the social bonds turned what was then a disparate collection of local networks into a UK movement with a national identity.

Back in 2012, organisations like Global Justice Now and War on Want needed strong local food justice movements with a global orientation to bolster their international campaigns. At the same time, there was a need for local food networks such as The Kindling Trust in Manchester, co-operatives like True Food in Reading, and growers like OrganicLea in London to join together and find a way to fix the system as a whole, rather than tackling their issues alone.

The Krems delegation utilised its connections and set about linking up these communities over the next few years. This new UK food sovereignty movement filled the political gap, while the LWA was still an emerging force.

The UK is unlike the land-based economies of France and Italy where rural movements are more common, so this movement had to make use of urban and peri-urban organisations to establish itself. It was imperative that people of colour and urban community organisations were part of the movement. An imperfect but earnestly meant intersectional process began, which saw larger in-person gatherings in London and then Hebden Bridge in 2015, alongside many mini gatherings across digital and physical space.

OrganicLea, the hosts of the first gathering in 2012, said that the event made 'picking through slug-damaged beans, on your knees, in the rain... feel like vital work: part of a movement, a wider struggle'. Attendees who had never met before found they were speaking the same language of food politics — making vital relationships which remain to this day.

It was these gatherings which inspired the next wave of mobilisations, and catalysed groups like LWA to form. Concepts like agroecology were road tested within UK parameters, and ideas

such as the campaign for a People's Food Policy were fostered. As the
LWA grew in strength the movement found itself with an effective
farmer-led institution, which effectively put food sovereignty and
agroecology on the map.

Each movement sows the seeds for the next. The politics evolve
and deepen to become relevant to the current zeitgeist. And it was
in the hot Austrian summer, around a table of beer and delicious
continental nourishment, that the LWA, and perhaps countless other
initiatives around Europe, were propagated.

DEE WOODS

Midnight Feast: A Joyful Celebration of Food and Farming and a Labour of Love

Landworkers' Alliances' Food Justice Policy Coordinator and cook Dee Woods provides a recipe (and music!) for gatherings

What started out as a late night barbecue of sausages accompanied by cider, chat and music at one of the Landworkers' Alliance AGMs, evolved to become a much loved highlight of the annual gathering of our members. Midnight Feast was a simple celebration of meat and bread washed down with lots of cider or beer, in the large kitchens of the venues we stayed at across the nations. It was a potluck of local ingredients, freshly baked sourdough bread, meat, herbs and vegetables, lovingly prepared throughout the day in between sessions.

Being nomadic in my work to build equity into our food systems, I bring back herbs and spices, or cooking influences, from the countries and communities I visit as a small way to share my experiences beyond policy advocacy, meetings and farm visits. A movement is built on our shared humanity, on the moments when we share joys and sorrows, when we sing and dance together; but most importantly, our bonds are woven and strengthened when we eat together.

The ultimate dish I would love to cook for the movement comes from my Caribbean Indigenous heritage. It is my own version based on oral history, taste memory, local ingredients and my skills as an intuitive cook, that I prepare mainly during winter. It is the epitome of slow food and a tribute to the fearless women of my bloodline who dared to dream, who took risks challenging the oppressive structures of patriarchy and racism, who cooked for their families, communities, and the political mavericks of the anticolonial movement who sought refuge in our family kitchen. Ancestors who walk with me as I take up the mantle of organising with the LWA for food sovereignty, climate justice, agroecological food systems, and good food for all.

For me, it aligns with the decolonial work of food sovereignty; the culturally relevant food integral to the right to food, and the reclamation of ancestral food ways that speak to the spiritual belonging to, and living in harmony with, the land and nature.

DEE WOODS

Dee Woods' Pepperpot Recipe

INGREDIENTS

Cured Brisket
- Beef brisket, 1.25kg
- Pickling spice (1 tbsp of each):
  - Whole peppercorns (white, red, green, black)
  - Pimento (allspice) berries
  - Mace
- 1 tsp dried chillies
- 1 head garlic
- 3 bay leaves
- 3 cinnamon sticks
- 3 cloves
- 2 thumbs length of fresh ginger
- 200g salt
- 75g sugar
- 2 litres water
- 1 lemon

Water
- 6–9 tbsp neutral tasting oil
- 2 tbsp brown sugar (optional)
- Salt and pepper to taste
- Caribbean hot pepper sauce to taste

Pepperpot
- Cured beef brisket, cut into large chunks
- 2kg of mixed wild game (rabbit, venison), stewing steak or oxtail, cut into large chunks
- 4–6 hot chillies (wiri wiri, scotch bonnets, habaneros), or more or less to taste, leave one whole and dice the others removing seeds and white membrane
- 500ml Cassareep
- 3 tsp roucou (annatto) powder or paste dissolved in water
- 6 cloves
- 3 cinnamon sticks

Music for this recipe:
- Freedom Music (Black Motion Aquarian Mix) by Irfan Rainy & Baba Israel

METHOD

For the Cured Brisket:
Start this recipe at least seven days before you want to eat it. Wash your brisket in some lemon juice and water. Set aside. Put all the brine ingredients together in a large pot and bring to a boil. Allow to cool. Use either a ceramic or glass bowl (or two large strong ziplock bags), place the meat flat inside and cover with the brine. Cover the bowl and place in the refrigerator for a minimum of 12 hours overnight and up to two days.

For the Pepperpot:
Heat the oil in a large oven proof cast iron or enamelled casserole. Add the meat in batches and ensure evenly browned. Transfer to a slow cooker if using or continue in the casserole. Add the chillies, cassareep, roucou, cloves and cinnamon. Mix, adding enough water to cover. Cover and bring to a boil then simmer until meat is tender and the sauce is black and glossy. Add the whole chilli to the top of the stew during the last 30 minutes of cooking. Remove the chilli and add salt, pepper, pepper sauce and sugar to taste.

For the Final Dish:
This is the hard part. Pepperpot tastes better after the flavours have been allowed to infuse for several days. It is safe to leave it unrefrigerated in a cool or cold kitchen as the cassareep and chillies preserve it, or you can put it in the refrigerator. But you must reheat it everyday!

After three days of doing this you can finally dive in. I normally wait a week. Pepperpot can be kept going for months or even years with the addition of fresh ingredients. Serve with flat cassava breads such as bammie, found in Caribbean food shops, or freshly baked bread. Enjoy with a glass of Guarap (wa- ra- po), a refreshing bubbly, fermented pineapple peel drink.

CHLOE BROADFIELD.

The Hope of Agroecology

Market gardener and agroecology practitioner Chloe Broadfield reflects on the potential of agroecology

We might sometimes forget that our fields are full of agencies and think that, as farmers, it is we who are imprinting our will upon the land. But in the thin space between the seasons, when all is shedding leaf and dying back, or else sprouting or budding anew, I see that we too are transformed through this relationship. Farming is as much about the cultivation of the human spirit, as it is about the cultivation of the land.

I work on the market garden at Tamarisk Farm in West Dorset, where the fields slope down to pebbled beach and sea. But this is not a job of the usual kind. It does not uphold the old structures, nor belong to the old economy. It is work re-imagined for a different path ahead; producing food, but also shaping ways of being in the world that have a hope of going on.

Agroecological farming is not the alienated wage labour that we close the door on with relief at the end of the day. It is work that flows seamlessly with life. The fields are a place of activity, but also of friendship and conviviality. In a day we might plant a hundred tomatoes, or sow a thousand lettuce seeds, but we might also pause in wonder at the goldfinches, or swim together in the sea and wash the earth from our knees.

To go home in the evening is rarely to leave the land behind. There is always an offering to carry away, and the connection with place spills out beyond the working day. We bake bread from wheat or rye grown and milled on the farm, we ferment and preserve the surplus fruit and vegetables, and we knit warmth for ourselves in winter from the fleece of the Tamarisk sheep. The culture that comes with farming is rooted in creativity and the skills for resilience, it has shown me the true abundance of life.

Equally it is a culture centred on care and community, and the farmers I have learnt from know that their community is more than merely human. Sheep, hens, cattle, and crops are as fundamental to daily life as we are to one another. So too are the wild species; the glow worms and mullein moths, and the ten different orchids that I have learnt to know by name. Working on the land unravels the modern myth that wellbeing is something belonging to the individual, and forces us to know, with the experience of each day, that it is rather the collective outcome of our relationships to one another, and to plant, creature, and climate.

What are these if not lessons that could carry us forward? I am hopeful for what the land might achieve through those who are moved by the meaning it holds, and who are stepping up to dedicate themselves to its care.

FRITHA WEST

A Lot Like Church

Late on Easter morning in a dusty shirt and boots
tramping by the river in the sun,
weaving past back gardens and out into the woods
the jays sounding alarm bells one by one,
as the piping oystercatchers and the twittering of tits
are interrupted by my own old Christian songs;
singing to my green cathedral with the only praise I know
— for when the feeling's true, the words are never wrong.
And the Spring springs up around me and the leaves are reaching out;
as I, a beggar to a saints' calm healing hands,
cry out all the Easter stories from the books and hymns and prayers
and this morning, I return them to the land.

OUT ON THE LAND

Statement of Solidarity with Trans Landworkers

The Landworkers' Alliance's LGBTQIA+ member led organising group Out On The Land offers their statement of solidarity with trans landworkers

As landworkers, we cherish the diversity and richness of the world we live in. Every living thing brings something beautiful to the ecosystem, each bringing a distinctive ingredient that makes it thrive. A woodland is made up of much more than just trees — from the birds flitting from forest floor to canopy, to worms working their way through the soil and the underground mycelial networks spanning the whole forest with their fragile threads. We see this in our human communities as well; old and young, different heritages and histories, neurodiversities, skills and tendencies. Our differences come together to make something bigger and richer than any one of us alone.

And as we fight to safeguard the beautiful diversity of our ecosystems — local and global — we fight to defend the diversity of our human communities. In both, that diversity is the foundation of our strength and resilience — the foundation of our survival. Because of this, we stand proudly and firmly with trans people; recognising that the struggle of trans people for life, dignity, and justice is our collective struggle to build the world we want to live in. The Landworkers' Alliance takes a stand against all forms of oppression, and from this principle it follows that we stand up for the right of trans people to live in a world free from violence, hatred, exclusion, and bigotry. We particularly reject the transphobia targeting trans women that seeks to contaminate the communities and movements that are dedicated to fighting oppression.

To all landworkers who are trans, nonbinary, and LGBTQIA+; we warmly welcome you as part of the Landworkers' Alliance community. We celebrate and value the presence of trans landworkers within the LWA, and will work to nurture allyship and solidarity among our cis-gendered members.

We also recognise that working for trans liberation is connected to resisting all forms of oppression and domination — white supremacy, homophobia, sexism, colonialism, ableism, and ageism. We are on a journey with building this understanding; we have a lot of learning to do and we don't always get it right. We strive to learn from our mistakes and not let the fear of getting things wrong prevent us from taking action.

The LWA is a complex ecosystem in itself, and there is strength and beauty in the breadth and richness of our membership. We know that this statement doesn't necessarily represent every single one of our members, and we are committed to having these conversations across our community: to listening and building common ground, to respecting each other as fellow walkers on a path towards food sovereignty, even when we don't see eye-to-eye.

As we walk this path we will centre the voices of those who are most impacted by the systems we are trying to dismantle, who have cultivated community, resistance and survival in its harsh, rocky soils. Within the LWA, we will support our working group for LGBTQIA+ members to create spaces, actions, and analysis that centres the experiences and perspectives of queer and trans landworkers. Part of this work is the Cultivating Justice project, a partnership between the LGBTQIA+ working group, LION (Land in Our Names), and Farmerama podcast, which will spotlight the stories of those most often marginalised and excluded from access to land and fair, dignified landwork.

So let's return to that forest at the edge of a lovingly planted field of crops. Smell the ripening fruit, dig your toes into the soft moss of the forest floor, listen to the countless creatures that pollinate the blossoms and turn the remains of plants and animals into the rich soil that feeds us. Know that the world around us is teeming with life, and in its beauty, none of it fits into the rigid and lifeless categories that strangle the will to create, to explore, and to survive. It is from this forest that we send this statement, a love letter to all those who, both now and in times past, trespass against the borders erected around private property, class, gender, the right to call a place home, and the ways that we love and move through the good earth.

—

This statement was released by the Landworkers' Alliance in early 2022. It was put out at a time when the fight for trans liberation faced struggles on multiple fronts: in healthcare, in the media, in politics, and in organising spaces. These struggles continue and sadly our own movement is not free from transphobia. It was in this context that trans landworkers and our allies asked the LWA to take a stand. Statements like this are important in laying claim to the politics of our movement. It is vital that we, in the LWA and the wider agroecological movement, speak up and stand up as, and in solidarity with, marginalised landworkers and people worldwide.

REBECCA LAUGHTON

The Power of Solidarity

The Landworkers' Alliances' Horticulture Campaigns Coordinator Rebecca Laughton on the forming of the LWA and the solidarity it provided land workers

Much of what I was involved in the years after graduating — working for the local food movement, living in the off-grid woodland community of Tinkers Bubble in Somerset, developing skills as a market gardener, and researching the 'human energy' aspect of land-based living — was part of agroecology, but I didn't think of it as such. Even at the first UK Food Sovereignty Gathering at OrganicLea in July 2012, where the Landworkers' Alliance was conceived, I was more concerned with finding a way to research the productivity of small farms; and had to leave early to run a food event in my local town of Haslemere. However, something began that day which, over the last ten years, has brought a focus and a sense of solidarity to what was previously a set of related but disparate individual projects.

The creation of the Landworkers Alliance (LWA) has its roots in a number of national and international movements. The classic pathway of many new entrants into farming is through volunteering on organic farms as WWOOFers (members of Worldwide Opportunities on Organic Farms), followed by finding work on a farm, and maybe then establishing their own business. For many years prior to LWA, Chapter 7, the planning branch of The Land is Ours, provided free advice to people trying to get planning permission for agricultural workers' dwellings, and in 2012 the Ecological Land Co-operative had already embarked on its path to create affordable smallholdings.[1] Many early members of LWA already identified themselves with the organic movement, but they struggled to gain access to land, acquire planning permission for new low-impact businesses, and develop the skills to establish a successful business.

From the start, LWA brought together a potent combination of activism and fun. AGMs have blended focussed discussions on CAP reform and access to land, with wild Fitty Gomash (aka The Rutted Furrows) cèilidhs and singing around the fire into the early hours. The sense of solidarity at my first AGM inspired me to help organise a protest outside Defra to mark the 2014 International Day of Peasant Struggles, where we created a pop-up farmers' market to demonstrate that small farmers have a role in food production too. There followed several years of creative mobilisations, ranging from Food Sovereignty Football in Parliament Square (you can see the film on the LWA website) to illustrate the unlevel playing field facing small scale farmers, to the Save our Standards pumpkin demo at the passing of the Trade Bill in 2020.

The activities of the LWA went up a gear following the Brexit vote of 2016. The opportunity to influence a brand-new agricultural policy focussed minds across the food movement; and generated a new level of coordinated activity between like-minded organisations. The Oxford Real Farming Conference has played a vital role in bringing the LWA together with the wider real food and farming movement, and the closing plenary provides us with an annual reminder that we are all part of something bigger. At this, and countless meetings throughout the year, the ever stronger connections with other organisations working on different parts of the same problem, and the friendships and sense of trust around shared purpose, are leading to a sense of growing confidence. The challenges we face — climate change, access to land, long working hours, and making fresh healthy food available to all for a realistic price — seem less daunting when viewed in the company of other activists and land workers.

For me, joining the LWA brought a powerful sense of solidarity. On returning to work after my first AGM, I remember wearing my new badge and thinking of all those other people working in the fields across the UK and feeling that I wasn't alone. I always feel nourished by meetings with fellow LWA staff or members; and energised for the challenges ahead. Those challenges may be great; but working with others towards the common aims of agroecology and food sovereignty is a more joyful journey because of companionship.

1. The Land is Ours is a land rights campaign set up in 1995, advocating for access to the land, its resources, and the planning processes. www.tlio.org.uk

DEE BUTTERLY AND MORGAN ODY

An International Movement of Peasants and Small Scale Farmers' Unions

Discussion between Dee Butterly of the Landworkers' Alliance and Morgan Ody of Confédération Paysanne and General Secretariat of La Via Campesina (LVC) on the role of LVC

DB   **La Via Campesina (LVC) is the international peasant and farmers movement that the Landworkers' Alliance is a member of. Could you explain what LVC is; why was it set up in the 1990s and what the political, social, and economic situation for farmers and peasants was at the time?**

MO   La Via Campesina is an international movement of peasants and small scale farmers' unions. It is an autonomous, pluralist, multicultural movement; political in its demand for social justice, but independent of any political party, economic or other type of affiliation. It has over 180 member union organisations in over 90 countries, representing millions of peasants, landless workers, indigenous people, pastoralists, fishers, migrant farmworkers, small and medium-size farmers, rural women, and peasant youth from around the world.

LVC was born in the 1990s in response to the process of

economic globalisation from finance institutions such as the International Monetary Fund (IMF) and the World Bank. These institutions were pushing for liberalisation and deregulation of the agriculture sector. At the same time the World Trade Organization (WTO) was pushing for globalisation of trade and inclusion of agricultural products in international trade, with governments saying that to ensure food security, we needed to open up all countries to international trade — making food cheaper and more accessible. This was felt by peasant organisations all over the world as a very strong threat to their livelihoods, as small scale farmers and peasants would be pushed to compete with cheaper products grown in other countries under different conditions of industrialised agriculture.

This was seen as the moment for small scale farmers and peasants to unify and organise; to push back against transnational corporations. If transnational corporations globalise the economy and trade, then we need to have a globalised movement fighting back. This is where the expression 'Globalize the struggle, Globalize hope!' came from.

DB **LVC was established in 1993, and now 30 years later there are millions of members of this movement world wide. Why do you think this trajectory of growth and power has been so effective?**

MO There are four main goals of LVC that have been very powerful and effective over the past three decades: first, to create unity in peasant organisations. We have a huge diversity of experience, understanding, ideology, and approach; which we cherish. It is key to make sure that small scale farmers, peasants, and agricultural workers around the world have mechanisms to understand and communicate with each other in order to create mutual understanding of our common struggles.

Secondly, is to ensure solidarity. When one member organisation has a problem — which happens very often, for example, facing repression from a right wing government —

then we show solidarity: we mobilise and organise through our media presence, protests, and building public pressure.

Thirdly, we make sure peasants and small scale farmers can represent themselves and their own voice in international institutions. Before LVC existed, agribusiness was represented on one side, and on the other there were NGOs talking on behalf of peasants. Our voices were not organised and we were not there at the negotiating table. We did not want to be directed by other organisations; we needed to be recognised by, for example, the United Nations Food and Agriculture Organisations (FAO), and speak on behalf of ourselves as peasants and small scale farmers.

And finally, the fourth goal is to develop joint campaigns against transnational corporations and international institutions such as the WTO, IMF, and the World Bank. These are really important common struggles and no single member organisation can challenge such big institutions alone. We cannot achieve change without being organised at an international level, we had to unify.

DB **The principles of food sovereignty are one of the core frameworks that LVC works for. What is food sovereignty?**

MO Food sovereignty as a framework was proposed by LVC in 1996 as a challenge to the framing of food security, advocating that the food system should be controlled by people and not corporations. Food sovereignty is defined as the right of peoples to healthy and culturally appropriate food produced through ecologically sustainable methods, and their right to define their own food and agriculture systems. We are fighting to keep farming, to keep being peasants, to keep producing food in an autonomous way. This is our aim, this is our struggle, and we need to protect ourselves. To ensure that the peasant way of life continues we need conditions, framing, and intentions to work from. The capitalist system aims to dismantle us. To keep being peasants, we need agrarian reform and access to productive resources such as land, water, and seeds. Once we have this we need to make decent revenue and business from our production, so we need prices that reflect the true cost of production. This is food sovereignty, and it supports the capacity for peasants around the world to produce food: we don't want to be fed by big corporations. We are able to produce food, and we want to protect the conditions to continue. To control our food system, we need not only to have fair conditions and markets, but to control how we produce. This is where the framing of 'agroecology' emerged.

DB **Different countries and movements interpret and put into action the principles of food sovereignty. Can you give an example of a recent process that worked to put food sovereignty into practice and action?**

MO The 2020–2021 Indian farmers' protest was precisely a struggle for food sovereignty. It was a fight to maintain strong public policies and regulation of the agricultural market and food system; to have fair prices and a decent revenue for farmers, and protect access to food in times of crises. These principles were strongly defended by farmers against the government and Indian food corporations. It was also the moment we witnessed a huge international mobilisation in solidarity with the Indian farmers' protest movement — solidarity with the situation in India, while recognising that these same issues are completely relevant here, and we have to organise against them in Europe.

DB **We are living in a time of great instability — politically, environmentally, and socially. What sort of strategy and direction do you think we should be focusing on at a local and international level in our organising — what does working at a collective level for change look like? Do you see challenges ahead, and what is the potential change we can make as part of an international movement?**

MO In the late 1990s and early 2000s, LVC was part of a very large and important anti-globalisation movement; there were many powerful forums and international gatherings such as the

World Social Forum, and we felt part of a strong international action for social justice. But in the late 2000s, there was a backlash; there was a lot of repression, and what was called 'the war against terrorism' created criminalisation. Many anti-globalisation movements were created in Latin America, but when right wing governments gained power huge problems were created that weakened many social movements.

One of the challenges now is to reconnect with other movements. We are not part of a big 'movement of movements' the way we used to be. We are facing huge humanitarian challenges and we need to create change. In order to do this, we need to rebuild alliances with other networks. Movements exist at the local level, and in some countries the national level, but at the global level, we do not currently have a powerful collective action. There are few big global actors and international movements that can create this change. LVC is in a good place, but how do we rebuild this collaboration?

We are not in an easy moment, we see the rise of the right in many countries. We have to find common understanding among movements from different places. It is not an easy job, but within LVC I feel that we are managing; that there is a peasant way to accept diversity and find compromise among different contexts and approaches, to find ways of maintaining unity. This peasant way is an approach where we can hold the complexity of multiple ideologies and approaches; and come up with strategies to navigate our movement collectively. The peasant way has a lot to do with patience. As peasants and small scale farmers we know patience; we know that first we have to plant the seeds and then we need to wait. This is critical, there is a right time for things and a wrong time. Working the land, working with the seasons, and working alongside animals helps us to develop a feeling for 'when is the right time'. As peasants, we have the capacity to feel what is happening and know the right moment.

DB **Can you explain the role of union members in LVC, how do we mobilise and build power at a grassroots level?**

MO   LVC is made up of member organisations. It does not exist if
      these members are not strong, organised, and participating.
      The reason we are here is to create unity and support for
      social justice. LVC supports members to challenge the issues
      they face. At LVC we believe that it's not a question of who is
      right or wrong; it is a question of the balance of power, about
      creating enough power that small scale farmers can oblige
      governments to change. For this to be effective in Europe,
      there needs to be an intersectional people's movement. Small
      scale peasants and farmers are not many, we need to organise
      as workers and unions. What we defend is much bigger than
      our own interests: we defend a vision of society, and of nature,
      that enables us to live in peace without domination. We need
      to build a collective action for social and ecological justice
      that is much bigger and more powerful than simply a peasant
      movement. Europe is the heart of capitalism and colonialism,
      it was born here. As organisations working at an international
      level, we need to politically frame and understand our history.

      There is not one model, there can be huge diversity in the
      struggle. We can bring together very different people who all
      share a common vision for social justice. Diversity is the best
      way to confuse those that we are organising against. There is
      no point opposing different strategies, the question is how we
      bring social movements together. Showing solidarity is key. It
      might start with climate justice or with organising for workers
      rights, with inflation and challenges to economic issues. We
      should be open and take part in other actions that we didn't
      create. At the same time we should be very cautious about who
      we build alliances with. This requires strong ideological clarity.

DB   **What is next for LVC?**

MO   What next? It is simple really. To keep being peasants, to
      maintain our farms and support other people to start farming
      and producing food. This is our foundation, it is not changing
      and it should not change.

Urban Abundance

CONOR O'KANE (TEKNOPEASANT)

Home Cooking

```
        C       F    C        G         C
Well come down my little darling, let's see what we got

        C    F  C         F      G
Besides a headful of dreams, and a big famine pot

        C      F    C               F G    Am
We've got seeds, we've got muscles, and what's left of our brains

      C   F    C          G         C
All we need now's a pushcart, and a couple of weans
```

It's time that we were growing, and not just for sale
We've got lettuce and leeks and twenty-two types of kale
We've got lovage, rosemary, basil and rue,
What we need are some parsnips and swedes for the stew

```
F     F    C      F C F C
Home, Home Cooking

G            F                 Am
Home Cooking, digging and delving with you
```

F
Home cooking,

G   C   G
Loving each mouthful of you

Now we don't need big acres, we don't need a farm,
We just need a spot to keep our seedlings warm
We'll tend them so gently through the cold days of spring,
Then we'll plant them out in summer, in the rain and the wind

Home, Home Cooking
Home Cooking, Digging and Delving with you
Home cooking,
Always tastes better with you

Then come mid-September with a wee bit of luck,
There'll be a fine harvest, some bounty to pluck
And we'll all sit around the big table, and we'll sing
Would ye look what ye get, when ye do your own thing

Home, Home Cooking
Home Cooking, Digging and Delving with you
Home cooking,
Loving each mouthful of you

JULIA LAWTON AND ANDY REDFEARN

A Decade of Growing Comes to Fruition

Brighton based CSA Fork and Dig It CIC share their response to the Covid–19 pandemic

We all experienced the shock of that first full lockdown during the Covid-19 pandemic in March 2020. At the time, we were feeling tremendously inspired having spoken with Satish Kumar, founder of Schumacher College, and read his book *Soil, Soul, Society: A New Trinity for Our Time*.[1] With permission from him, *Soil, Soul, Society* became our maxim as we reached out to our local community with the following initiatives.

CROP HAMPERS — May 2020
Hamper with 12 crop plants delivered to 35 family homes and gardens, inspiring people to grow their own.

> The Covid-19 crisis has shown how vulnerable our food system is. It has highlighted the need for a transition to agroecological farming and food sovereignty. Supermarket shortages and issues in global supply chains demonstrate the uncertainties of our food system, which relies on imports and migrant labour.
>
> Our strong, core team of trainees and long standing volunteers are onsite every week growing, sowing, transplanting crops

and most importantly, continually building soil structure and fertility. Small scale CSAs nationwide are responding to the increase in demand for fresh, locally produced organic food.

Society and government can lead a transition towards a food system that ensures everyone has access to healthy affordable food; a system that values workers and producers, and has a local supply chain. Sustainable agroecological farming principles will ensure that our farming methods contribute positively to the climate and biodiversity crises that we face.

We want to encourage and inspire you to grow your own produce. To do this we have put together a garden 'Crop Hamper' of 12 different crops. The plants in the hamper have been raised from seed, grown and hardened-off to maximise your success. They are strong hardy plants.

The Crop Hampers will be ready for delivery in mid May (after the risk of frost has gone) and include: tomatoes, peppers, courgettes, pumpkins, strawberries, spinach, chard, kale, parsley, sage, french beans, lettuce, and red/ black currants. To ensure success you will also receive instructions, a bag of compost and a comfrey plant to enable you to make your own fertilising tea.

No food miles, harvested fresh when you need them, bursting with flavour and nutrition.

KICKSTART GARDEN HAMPERS — August 2020

Over 1000 plants delivered to 110 households to grow on their own leafy greens throughout autumn, winter and spring.

Many thanks for supporting Brighton Community Supported Agriculture by accepting our Kickstart Garden Hamper. We simply cannot grow enough organic food on our two acre site on Stanmer Organics, Stanmer Park and are delighted that you have decided to embark on the fulfilling journey of starting and continuing to grow your own produce.

It is our mission to inspire, encourage and support more and more people to grow as much decent food as they can. We firmly believe that locally produced food is the best way to go towards securing a better food system for all. Good for our health and the planet's health :).

Organically grown from our mostly saved seed, these plants will produce nutrient dense food, full of flavour, texture and colour. Bursting with antioxidants, vitamins and minerals they will provide you with fresh, tasty greens from your very own small market garden.

Organic food with no nasty chemicals, no carbon footprint, no supply chain and harvested when you want so no need for any packaging either :). By observing and harvesting when the crops are just right for picking you can either eat straight away, give to a neighbour, or preserve for another day by pickling, drying or blanching and freezing — helping to further reduce food waste and encourage true seasonal eating.

You have received nine plants — three spinach, three chard, three kale; and instructions on how to successfully grow those plants on. The leaves of each crop catch the sunlight and change it into food/ energy for the plant to use. We then eat the crop and release all the goodness (nutrients, vitamins, minerals) and energy for our own use — maintaining healthy bodies, repairing damage, renewing cells and organs and basically fuelling our body, mind and soul.

It's like we are eating the sun's renewable energy whilst recycling. And so, over to you… observe, enjoy, harvest. We wish you a successful growing season ahead :). Soil, Soul, Society.

KICK START COMMUNITY GARDENS — August 2020
1500 seedlings delivered to six community gardens around the city to grow on to produce months of leafy greens throughout the winter.

DONATIONS TO LOCAL FOOD BANKS — June-July 2020
An extra row was grown for FareShare and The Real Junk Food

Project, who intercept and use waste food. 50kg of leafy greens harvested and delivered to add to food parcels or community meals.

SUPERGREEN BOOST BOUQUETS — July-August 2020
150 Care Home staff received a 400g bouquet of spinach, chard and kale leaves to say thank you for their hard work the past few months.

> We recognise you have been put under unimaginable stress and strain over the past few months by the Covid-19 outbreak. You have continually put yourselves in the firing line, struggled to get PPE, been undervalued, underpaid and overlooked.
>
> We wanted to say a big THANK YOU for the work that you do and present you with a boost of locally grown, organic, fresh produce. Full of antioxidants and nutrients, good for your health and your immune system to look after you, for once, during these challenging times. Best wishes from Our Covid-19 Response Team. Soil, Soul, Society.

—

2020 was our best growing season to date. It was the most food we have ever grown, with the most people we had had onsite; and our organic fruit, vegetable and seedlings reached more people than ever before!

Money didn't power any of the initiatives. Good will, good leadership and community coming together did. We created a circular economy powered by love. The feelings, observations and reflections from this unusual growing season continually remind us of what we can positively achieve together.

1.   Kumar, Satish. *Soil, Soul, Society: A New Trinity for Our Time.* Brighton: Ivy Press, 2013.

RU LITHERLAND

Greetings From the Smoke

Grower Ru Litherland from OrganicLea writes about the parallel and interconnected journeys of LWA and OrganicLea

As the Landworkers' Alliance celebrates its tenth birthday, we at OrganicLea in East London, mark our 21st. It has been fun and instructive to reflect on how the Landworkers' Alliance's increasing dynamism and impact at a national level, is reflected at a local level by a host of grassroots (perhaps that should be leguminous roots!) organisations and networks. Here is our potted-up history as one illustrative example.

OrganicLea was born at the turn of the century, when the anti-globalisation movement had confronted, but not derailed, the World Trade Organization (WTO) and other global capitalist players. Some in the movement were calling for something akin to a 'long march through the communities' to create change from the ground up. At the same time, a series of food scandals emerged — salmonella, 'mad cow disease', foot and mouth, plus genetic modification (GM) — repeatedly highlighting the fact that our industrial food system was broken.

We established OrganicLea as a workers' co-operative, feeling then, as we feel today, that food is a good place to start working for social

and economic justice. A local food growing scheme provides myriad
social, environmental, and economic health benefits and cultural
connections. We also felt that the Lea Valley, an area with a rich
history of food growing and market gardening, would provide a
meaningful base for such a scheme. Here is how we grew.

2000 — First project ideas meeting and food-based fundraising
benefits. Over 100 letters sent to local authorities and organisations
in the Lea Valley. The London Borough of Waltham Forest's
Allotments Officer sent the only reply — offering six derelict
allotment plots in North Chingford.

2001 — The first soil was turned in OrganicLea's name, with bramble
clearance, raised bed-building, and a shed donated by the local
Resident Association. Weekly volunteer days and informal training
sessions on the allotments brought people together to learn and share
the harvest. We grew food and built skills, knowledge, and connections
on the peri-urban allotment plot, and it became a springboard to

reach the heart of urban communities with stalls, events, and growing projects in places such as housing estates and schools.

2003–04 — The first organic gardening training courses at the allotment. We organised a meeting in Walthamstow with 50 participants from local community and health groups discussing how residents could get better food, and the idea of establishing a food hub was developed. The local Hornbeam Café became this hub with a weekly market stall supported by Eostre Organics growers' co-operative in East Anglia. The café mopped up unsold produce, and hosted other food events, workshops, and info-sharing. Eostre and Hackney-based Growing Communities supported us to start a veg box scheme.

2007 — OrganicLea publishes *Selling Allotment Produce — Is it Legal? Is it Right?*, which attracts some interest in policy circles. On the back of this we launched our Cropshare scheme, enabling home and allotment gardeners to share or trade their surplus fruit and veg through our market stalls.

2008 — OrganicLea growers began to tutor City & Guilds-accredited horticulture courses with Waltham Forest Adult Learning Service. We also helped with community planting of a linear orchard along a cycle path linking Walthamstow and Chingford. When collecting the trees from the council's Hawkwood Plant Nursery, we were told it was about to close. Appalled by the idea of half an acre of glasshouse being demolished, we wrote to the council suggesting we turn it into a community market garden and plant nursery. A year later, we were offered a trial year.

2009 — The council agreed to a ten year lease and we secured Lottery funding. *Hawkwood Nursery: for Plants and People — A Journey in Permaculture Design* was published. Now, we were a ten-person co-operative composed largely of waged workers, and had scaled up production, plus training and volunteering work. The first of over 200 trees were planted, and most of the council's horticulture classes were relocated to Hawkwood. We also moved our box scheme, which grew steadily in the following years fuelled by produce from our East Anglian friends and our own market garden.

We developed weekly trade with pubs and restaurants in the capital. The site has evolved organically, imprinted by all the people who have passed through.

2012 — We were proud to host Transforming Our Food System, a gathering attended by 110 food campaigners, which helped to launch the Food Sovereignty movement in the UK and kickstart the formation of the Landworkers' Alliance.

2010–2021 — Since the dawning of our 'Hawkwood Years' and the emergence of the LWA as a force, we have grown our box scheme to over 1,000 veg and fruit bags per week; grown 9,833kg of organic produce in the last year alone; and now work with over 300 volunteers at the community market garden every year. Last year we enrolled over 100 learners onto our accredited courses; and ran an apprenticeship, traineeships, and programmes for young people with special educational needs. We established a further two weekly market stalls in East London (which sadly closed in 2021 so we could concentrate on our box scheme). We have been directly involved in helping to establish over 50 food gardens with community partners, including schools, housing associations, community centres, and supported housing gardens, in Waltham Forest and neighbouring boroughs.

—

We are proud to operate as a workers' co-operative and now have 25 members. More recently we have been exploring how we can better represent the diverse community of Waltham Forest within our co-op, and actively challenge structural racism and other forms of oppression.

The ideas and concerns that were a catalyst for the formation of OrganicLea in 2001 were fringe at the time, but now have become more widely accepted. Too often the 'little people', and initiatives like OrganicLea, are regarded as beneficiaries of the zeitgeist, but we should remember we are also its creators.

We have worked creatively to change opinions and culture from the bottom up. We have built systems, structures, and ways of working that practically demonstrate a better way of working with the land

and each other. This gives weight to the LWA's push for a policy environment that supports such efforts. And when push comes to shove — as it did during the Covid-19 pandemic — we can safely, graciously fall back on the work we have done and the structures we created. We can rely on it all to feed both our bodies and souls.

ROBIN GREY

The Ballad of Hawkwood

```
        C                    C      F
There is a fine gent christened Ru Litherland

C                    F
Mulch, sow and then reap

        C                    C
There is a fine gent christened Ru Litherland

        C                    C
And he has green fingers on both of his hands

C G/b C        F        C        G      C
I'll be   good to the land and the land will be good to me
```

With a co-op of comrades he dreamed a bold dream
To grow food for his kinsfolk as nature decreed.

By the edge the forest they spied a fine patch
And to grow fruit and veg there a plan they did hatch.

But the men of the hour dreamed of buildings not plants
A development would far more their profits enhance.

Our ancestors fought for this fair forest land
So now against the law was the businessman's plan.

After twelve months had past did the council relent
Now we'll work the earth as our ancestors meant.

Now if you pass by here you might hear a tune:
Mulch, sow and then reap
Now if you pass by here you might hear a tune,
The melody is old and the words will be soon.
I'll be good to the land and the land will be good to me

—

(cc)

ROBYN ELLIS

Bringing the Land Back into the Cities

Permaculture grower Robyn Ellis discusses urban growing in Hulme, Manchester

A defining moment in my land work journey was realising how we can find our relationship with the land when we live in concrete jungles. When instead of acres to work with, we have a balcony, windowsill, or front doorstep as our only outdoor space. This is why our journey back to the land is what occupies our world here in Manchester. As a grassroots permaculture project, The Gaskell Garden Project draws on the patterns found in nature, and takes responsibility for our existence through using sustainable low energy methods to increase biodiversity and widen access to growing food in the city.

But how do we rekindle this ancient connection between people and soil and the autonomy of growing our own food, when we have been separated for so long? When we have restricted access to green spaces and limited financial means to move elsewhere? When our green spaces are simply built upon by the highest bidder, blocking out our sun. A defining moment was being granted permission to begin a community asset transfer of 1.2 acres of unused (except as a frequent cycle path, cut-through, and home to local parakeets!) land in Hulme.

The land is next to the busiest road and most polluted area in Manchester. The 1.2 acres consists of an old slip road, tarmac, and a small patch of grass. We are trying to find new ways to re-learn the relationships between ourselves and our land after generations of separation, transforming wasted land into a possibility of a biodiverse green community space and edible forest. There has been interest in exploring the benefits of using disused urban space for growing food, however, it is important to look at how many of these spaces are being transformed in the name of progress.[1] Many 'meanwhile projects' by external organisations are separate from the communities living around them and consequently, are not inclusive of the people and the space they are in. Because of this, they are damaged, and not sustained by the community.

Creating a sense of ownership towards land in a city is a challenge. The concentration of people living on top of one another means the options for space are limited. The constant re-development of land for unaffordable housing moves long-term residents away, and can create a fractured sense of belonging to space. In Manchester, in light of the pandemic, communities are taking charge of their spaces for their needs. From back alleyway projects to community gardens in parks, people are redistributing food from front doorsteps, and transforming disused spaces into something from which communities can thrive. There is something inspirational about change coming from the ground upwards, but it also feels that, considering the negligence from those in power, there is no other option to make tangible changes for our futures.

The impact of concentrated urban living is being experienced in the quality of air we breathe: the consequence of importing fruit and veg from outside of the UK is that carbon in cities is reaching irreversible levels. The more city households and community green spaces dedicated to growing food for urban neighbourhoods, the more biodiversity, carbon reduction, and pathways towards equal access to food. Bringing people back to land is a lesson we are relearning from the climate crisis, but bringing the land back into cities will be integral in this shift too.

1.   Harvey, Fiona. 'UK could grow up to 40% of its own fruit and vegetables by using urban green spaces.' *Guardian*. 24 Jan 2022.

MIM BLACK

Edinburgh's Anarcho-Mycologist Workers Cooperative: Rhyze Mushrooms

*Founding member of Rhyze Mushrooms Co-op
Mim Black shares the beginnings of a radical
community mushroom farm and education
project in central Edinburgh*

In May 2021, seven of us pooled some cash, bought a second-hand shipping container, and placed it on an asbestos-ridden brownfield site in central Edinburgh. After the sudden halt in the climate organising on the streets, which had originally brought us together, we had spent 2020 on Zoom dreaming and scheming up a project that sought to build community resilience through urban food production. We had become disillusioned with asking the people in power to do the right thing, we felt it was time to start building the world we wanted to see where we stood. So, we decided to start a community mushroom farm.

We are Rhyze Mushrooms, a community mushroom farm and education project based in Edinburgh. We are helping to build local food sovereignty and community resilience by growing nutritious food in disused urban spaces, and teaching others how to do the same. Taking our cue from the nutrient cycling mushrooms we grow, we like to say that we are trying to decompose the corporate food regime.

There are no silver bullets or panaceas — we know that we need a systematic overhaul of food growing practices and the power relations that uphold our current unjust system. However, we do believe that mushroom cultivation should be taken seriously as an important tool in building agroecological food systems.

Mushrooms are a highly nutritious crop that can be grown seasonally outdoors, or year round indoors. Indoor mushroom cultivation in particular is very resilient to climatic change, and with super short cropping cycles (about four weeks), it can deliver fresh food at times of the year when harvests are otherwise slim. As nutrient cyclers they can be integrated into growing systems to build soil for horticulture or remediate runoff from animal husbandry.

Mushrooms also thrive in places where plants do not — sustainable local calories that don't add pressure to scarce arable land is no small thing. Indeed mushrooms can thrive in marginal spaces that are actually quite abundant: car parks, contaminated sites, basements, and dark sheds. More than half of the world's population now live in cities. Modes of food production that can be integrated into urban landscapes — bringing people closer to nature, and putting food production into community hands — need to be taken seriously.

Beyond these practical benefits, we believe in the discursive, imaginative, and transformative potential of mushrooms: through engaging with fungi we can re-enchant our understanding of this ecosystem that we humans are just a small part of; fungi are magical, fungi are queer, they are community organisers, and alchemists.

Throughout 2021, using scavenged and gifted materials, and lots of elbow grease, we worked to retrofit our now bright yellow shipping container into a farm that can be plugged in anywhere. We are currently based on a meanwhile site which is slated for development into flats next year, which comes with challenges and benefits: we don't have access to running water, so we collect and filter rainwater, but we also don't have to pay rent!

We have become self-taught mycologists growing about 40kg of gourmet mushrooms a month. We grow different edible mushrooms like oysters and lion's mane primarily on urban waste that we collect from local businesses, such as hardwood sawdust from carpenters and coffee chaff from coffee roasters. We sell our mushrooms to local greengrocers wholesale, and we sell direct at the farm door and at farmers' markets.

We are constantly improving our process and experimenting with different methods and substrates, with the aim of eventually getting to a point where we have a really solid, completely circular growing process that uses innovative methods and waste streams to produce food.

Education and community building is central to our work. We run workshops for individuals and community groups, teaching people how to grow mushrooms at home and in their community spaces. We also like to put on a good party — keep an eye out and come on down.

Whose Land?
Our Land

SAM LEE

You Subjects of England

You subjects of England come give this some time
A story I'll tell you of what's soon to be crime
New laws that will take you from knowing your land
Like old we are pressed to defend where we stand

If rules are made up to deny what should be free
To prosecute those who to walk is to breathe
That rivers and fields that are crossed will mean time
Does England reveal a renewed rule of unkind

So come all you subjects and question your rights
Do we accept such attempts or stand for our rights
Will the rocks melt and the seas start to boil
Generations to come will fear their own soil
You ramblers, travelers, wild swimmers and all
We stand all as one or down we will fall

—

*Song lyrics by Sam Lee, commissioned by Nick Hayes
for Right to Roam and Drive2Survive.*

JACKIE BRIDGEN

Stand Up for Tenants!

Long-term tenant farmer Jackie Bridgen from Chestnuts Farm, a smallholding and CSA project in Wiltshire, makes the case for supporting tenant farmers

I would like to take a stand for tenants.

Chestnuts Farm has been a member of the Landworkers' Alliance since around 2015, and has seen it grow and change into a powerful social movement to be reckoned with. During that time we have also seen access to land become more problematic (not to mention expensive) and the barrier that many new entrants face — that is, difficulties in buying or getting access to land — has been raised higher.

Post-pandemic, the trend for wealthy city dwellers to buy up a few acres and work from home in a rural setting has only made matters worse.

It can be overwhelming to see others succeed in their land-based journeys because they have privileges or advantages, which others simply cannot access. They may have wealthy parents willing to invest, or even a family farm to inherit. They may even have certain skill sets you have as yet been unable to obtain — social media skills for example. Or they may simply have no ties, enabling them to obtain land and camp out for the summer, making a start on their business. Meanwhile, with the rent to pay and kids to feed, passionate, talented people despair of ever getting their own plot.

Renting has got bad press lately, and to some extent, rightly so. Since 1995 when the Farm Business Tenancy (FBT) replaced the Agricultural Holding Act Tenancy, terms have gotten shorter, despite the good work of the Tenant Farmers' Association in campaigning for minimum ten-year tenure to enable the tenant to plan ahead and build a business. If tenancies are short and insecure farmers feel unable to invest. In many cases the farm is a home as well as a workplace, and so the household is also subjected to this insecurity.

However, inspired and fired up new (and not so new) entrants around the globe have created barnstorming businesses with nothing approaching the security of even the shortest of FBTs. Information now abounds about how to navigate lack of tenure security. The work of pioneering farmers, Joel Salatin and 'The Urban Farmer' Curtis Stone to name but two, demonstrates the flexibility of purchasing only 'infrastructure you can take with you' — electric fences, water tanks, trailer based operating facilities, 'chicken tractors' — in the terrifyingly unregulated USA.

Farmers and peasants around the world are engaged in land rights struggles, campaigns and direct action, while living with the precarious nature of borrowed land. They are succeeding in feeding their communities against the odds.

We came into farming in our 40s, late for new entrants, and were not in a position to buy land. We rented our land on a handshake at first, and were therefore reluctant (not to mention unable!) to invest much in the way of infrastructure. Over the years we negotiated with the then tenant, the agent, and the ultimate landlords (Crown Estate); and managed to get a proper FBT, first for five and then renewed at ten years.

There is a tradition and a history of tenant farmers in this country, but this is threatened by county councils selling off farms, which were originally gifted to them for the purposes of encouraging new entrants, and by bulk land purchases by corporate investors.

While good work is taking place to enable wider land ownership, for example by the Ecological Land Cooperative, please let us reclaim tenancy, and make it the force for good it can be.

I passionately believe that we need to support new tenants through policies that incentivise land owners to provide long-term secure tenancies, and which think creatively about tenancies and how they can provide a better experience for tenants. We need to change the face of tenancy, and make use of every acre we can get hold of to create the food sovereignty and local food security we believe in.

JOSINA CALLISTE

Why We Need Racial Justice in Farming!

Taken from her 2020 blog post, co-founder of Land in Our Names (LION), Josina Calliste, writes a personal reflection on the need for racial justice in farming

I am an alumna of Soul Fire's farming immersion — a heavily-oversubscribed six day programme designed for Black, Indigenous and Latinx people 'to gain basic skills in regenerative farming in a culturally relevant, supportive, and joyful environment'. Prior to going to Soul Fire, I would spend a lot of time doing permaculture courses in Britain and Europe. One could say, 'There are so many places to learn about food production closer to home — what was so special about this particular course?'

Having taken introductory permaculture courses, participated in 'Permablitzes', and spent a month on a food forest course in Portugal, I was experienced in being in environments where racial justice was rarely mentioned; all teachers were white, and there was zero mention of African or indigenous farming practices which influenced modern, regenerative or sustainable farming practices. I had previously felt that I was screaming into the void when it came to recognising the roots of oppression or racial injustice in these spaces.

I co-founded Land in Our Names (LION) in June 2019, two months before going to the USA. There is a growing movement of BPoC (Black and People of Colour) bringing an anti-racist lens to nature connection, farming and justice movements focussed on food, land and climate. Although the context for anti-racist activism differs greatly between these different settings, we are learning so much about racial justice from our American cousins. So it felt important to tap into the common understanding shared at Soul Fire Farm, especially for my growth and development as someone new to land activism.

I was lucky to have the income and time off work to be able to afford this trip, but still questioned spending money going all that way — for a mere six days farming? (Programme rates are offered on a sliding scale, but flights are still pricey). I further justified the cost by padding out the trip with family visits across the stolen lands in 'New England' and a delightful schoolfriend's circus-themed wedding. However, I figured that no one else I knew in the UK had done this course, and there were plenty of learnings I could bring back home to share with others, especially among people who wouldn't be able to afford it. Leah Penniman from Soul Fire Farm and I chatted about her upcoming trip to England to give the keynote at the Oxford Real Farming Conference in January. It felt like the start of something, building a movement that was simultaneously overdue, but that no one was really prepared for.

My urge to go to Soul Fire Farm was proven justified. It was profoundly life-changing for a black woman interested in farming to have our African ancestors' stories of survival and resistance interwoven with classes on how to grow her own food. The open acknowledgement of enslavement as a site of pain is highly relevant for many people in the African diaspora. Especially ones like me, whose families have farming or land work as part of their histories of exploitation and violence.

Upon returning, I held a dinner, where I shared many of my learnings from Soul Fire; attendees were black and brown friends who remain involved or supportive of LION's work. It felt clearer to proudly acknowledge what our task is; to heal ourselves and seek land reparations.

There are millions of BPoC cultivating plants and growing food. Our voices, faces, and wisdom are not heard enough in food growing spaces. LION is a young collective, still in the process of seeking out the voices of elders who come before us, developing healthy working practices and making sure we connect to the right communities. Many of the Black perspectives on land I want to learn from are hard to find. There is never enough time to read, so as a new land activist, it's super hard to get rid of the imposter syndrome.

In January 2020, I started working on LION full-time. I had previously worked in sexual health and for 18 months had my dream job with a small group of queer activists. However, I felt that in the field overall, I had learned everything I had wanted to learn. Many people may be questioning whether their job is not serving the immediate needs of our living planet and for me, I knew I had to commit my life to learning how to grow food, and expanding land access for everyone. Knowing that my labour contributes to hope of a better future helps fight feelings of desolation, but also means I'm reminded what we're up against every damn day.

Anyone doing racial justice work has to avoid being a 'self-narrating zoo exhibit'. We will be asked for free consultations on diversity and inclusion by organisations who have not committed resources to anti-oppression in the past. Panic ensued after the latest Black Lives Matter news cycle in June, with many organisations anxious of new scrutiny in their recruiting and working practices. In June 2020, the deluge of requests filling our inboxes jarred with the imposter syndrome I felt. On one hand, I was heartened that people contacted us to ask how they can support us through donations and redistributions, yet annoyed that the much-needed racial justice work had been so chronically underfunded that a young collective like LION could be seen as 'existing leaders' in the field.

We do not turn down all requests to speak to majority-white audiences but it's not our central goal. We did not start to become diversity consultants, and we cannot shortcut the long-term work of building anti-oppressive working practices in white spaces. We want to tell BPoC communities in Britain of our dreams, of land for common good, in our (collective) names. What would a land movement in Britain look like if Afro-indigenous ancestral farming practices were at its heart? If people writing food policies were from the communities most affected by food inequalities? For fighting industrial agriculture, black communities need land power. To achieve climate justice, people of colour need to be able to grow their own food. For environmental justice, the spiritual ecologies of indigenous communities should be at the front and centre.

In writing this I'm reminded of the magic of the Soul Fire programmes, and encourage white readers to avoid all notions of a paternalistic 'how can we help them?' Instead, think of how you can support and resource the movement to have our own black-owned farms here. Our goal shouldn't just be trying to get black and brown faces into white places, but uplifting the visionaries who can lead conversations in their own communities. LION's events with BPoC communities ground us in our purpose. The post-Oxford Real Farming Conference 2020 caucus was life-changing for me, and many participants echoed these sentiments. I cannot tell you how important it is that we hear from each other on our own terms, and augment the functions of food producing spaces to encompass political and historical learning and organising.

To battle imposter syndrome, I have a screenshot from the Soul Fire Farm instagram page as the wallpaper on my phone. It's a photo of me and Olá (the other LION co-founder) holding a fluffy chick at Willowbrook.

Underneath, the text reads:

> Today was historic! 40+ UK Black & Brown farmers gathered at @willowbrookfarmers convened by @landinournames[...] We are rooting for Josina, a #soulfirefarm alum, Ola, and the other visionaries catalyzing this land based movement.

Of course it's not affordable or sustainable for all aspiring BIPoC growers to go to Soul Fire Farm to learn how to build an equivalent movement and farming spaces. So we need all the support we can get to build the movement locally. In the spirit of 'a high tide raises all ships', racial justice in the progressive farming movement will benefit everyone.

ALEX HEFFRON AND KAI HERON

An Agroecological Revolution

An excerpt from 'Renewing the Land Question: Against Greengrabbing and Green Colonialism,' first published in New Socialist on 20th February 2022

It is possible to mitigate global heating, avert ecological collapse, and address historic land injustices if we're willing to entertain radically different patterns of landownership and land-use. For decades, producers and peasants in the Global South, led by La Via Campesina and others, have argued for food sovereignty and agroecological farming as a means to bring people together in internationalist, anti-capitalist solidarity. Cuba is leading the way with its *organipónicos* (urban organic garden) cooperatives.[1] More than half the area of the Havana Province is under cultivation, with urban and peri-urban farms supplying 70% of its vegetable needs,[2] 10.5 million litres of milk and 1,700 tons of meat.[3]

Although 70% of the land in Britain is used for agricultural purposes, the country produces just 55% of what it consumes.[4] Of that 55%, much is grown and harvested, cared for, and slaughtered by seasonal, casualised, gendered, and racialised immigrant labour; under brutally exploitative conditions. The rest of Britain's food is either grown under similarly exploitative systems in the EU's industrialised and polluting agribusiness sites, or in tropical climates where super-profits can be secured through the super-exploitation of the Global South's lands and labour. Food production accounts for 34% of the world's greenhouse gas emissions, and almost a third of agricultural emissions are due to food storage and distribution.[5]

Food sovereignty demands localised food production, and for populations to have access to food that is affordable and culturally appropriate. It moves us away from productivism and centralises social reproduction, dislodging capital and empowering peasants and women in the process. Agroecology calls for more people to work on the land, growing nutritious food to meet human needs whilst restoring ecological flows and biodiversity. It desires a blurring of the lines of demarcation between leisure and work, country and city.[6] It is in favour of localised democratic management of land, set within the clear objectives of increasing food production without dependence on agribusiness, while tackling food insecurity and reconnecting people with where their food comes from. When combined with more efficient land use,[7] such as agroforestry, the cessation of biofuel production, and a shift in diets towards plant-based proteins, pasture-fed livestock, and seasonal fruits, nuts and vegetables; agroecology could feed Europe.[8] It can also play an important role in sequestering carbon in soil and biomass, reducing overall greenhouse gas emissions from agriculture by around 40%, and increasing resilience to climate chaos.[9] As Max Ajl writes, wresting back control of food and farming allows us an 'entry point into restructuring our world.'[10] As we've argued elsewhere, it allows us to provide for ourselves beyond capital's circuits of reproduction. Moreover, through agroecology, as Chitaya says, 'women can teach men, Black people can teach white people, the poor can teach the rich.'[11]

To achieve this shift, the question of nationalising the land, towards decommodification, is essential. To prevent further corporate takeover, to grant land access to historically excluded groups, to ensure land is used most effectively to reverse ecological breakdown and avert further climate collapse, the land must be put under democratic ownership. To succeed, land nationalisation must be facilitated through local democratic control within a broader national framework, with goals such as rapid decarbonisation and the reversal of biodiversity loss at its centre.

To begin, a national land bank could facilitate the purchase of new land that comes onto the market, to let out to tenants at a low rent, to farm in accordance with agroecological principles. This could be achieved through the reinvigoration and re-imagination of the county farms estate, enabling the interplay between local and national aims. County farms were established during the late-Victorian agricultural depression with the aim of easing access to land for those who couldn't afford to buy land or pay rent at market rates. Today, these farms are being sold off at an alarming rate, as councils seek to pay for essential services and reduce deficits. Shrubsole has found that the extent of county farms in England has halved in 40 years.[12]

This short-term thinking is destroying an organisational form that could serve to undermine landlordism and build the basis for a new commons. Food produced by these farms could be purchased by the public sector through localised public procurement platforms, as has been considered in Wales. Land that was appropriated by the military, such as the common lands of Epynt, near Brecon, could be returned to communities and land workers to manage in common. The Crown Estate — all 106,000 hectares of it — is also ripe for expropriation. These are not the limits of our horizon, but they may go some way towards establishing agroecological communes and rebuilding the left's relation to land and the material basis of our society, whilst undermining capital's control of the countryside.

Speaking in 1872, Marx knew that revolutionary land reform would bring about a dramatic shift in the balance of forces between the working classes, landlords, capital, and the state:

> The nationalisation of land will work a complete change in the relations between labour and capital, and finally, do away with the capitalist form of production, whether industrial or rural. Then class distinctions and privileges will disappear together with the economical basis upon which they rest. To live on other people's labour will become a thing of the past. There will be no longer any government or state power, distinct from society itself! Agriculture, mining, manufacture, in one word, all branches of production, will gradually be organised in the most adequate

manner. National centralisation of the means of production will become the national basis of a society composed of associations of free and equal producers, carrying on the social business on a common and rational plan. Such is the humanitarian goal to which the great economic movement of the 19th century is tending.[13]

Little essential has changed. Then, as now, the power of state, capital and landowners to decide how the land is used rests only on what the 18th century poet and opponent of enclosure, John Clare, called 'the lawless law' of racialised violence and expropriation. It is time to renew the land question and rebuild the commons.[14]

1. Ratchford, Aidan. 'Agroecology and the survival of Cuban Socialism.' *New Socialist*. 16 October 2021. <https://newsocialist.org.uk/agroecology-and-cuban-socialism>

2. Altieri, Miguel and Nicholls, Clara. 'Pathways for the amplification of agroecology.' *Agroecology and sustainable food systems. 42:10*, 2018. pp1170–1193.

3. 'Growing greener cities — Havana.' *Food and Agriculture Organization of the United Nations*, 2014.

4. 'Food Statistics in your Pocket.' *Department for Environment, Food & Rural Affairs*. 7 November 2022. <https://www.gov.uk/government/statistics/food-statistics-pocketbook/food-statistics-in-your-pocket>

5. 'Food systems account for over one-third of global greenhouse gas emissions.' *UN News*. 9 March 2021. 7 November 2022. <https://news.un.org/en/story/2021/03/1086822>

6. Heffron, Alex and Heron, Kai. 'Towards the Abolition of the Hinterland.' *Architectural Design*. 92:1, 2022. pp120–127.

7. Billen, Gilles. et al. 'Reshaping the European agro-food system and closing its nitrogen cycle: The potential of combining dietary change, agroecology, and circularity.' *One Earth*. 4:6, 2021. pp839–850.

8. 'An agroecological Europe in 2050: multifunctional agriculture for healthy eating.' *IDDRI*. 2018.

9. Altieri, Miguel et al. 'Agroecology and the design of a climate change-resilient farming system.' *Agronomy for Sustainable Development*. 2015. pp869–890.

10. Ajl, Max. *A People's Green New Deal*. London: Pluto Press, 2021. p126.

11. Patel, Raj. 'Agroecology is the Solution to World Hunger.' *Scientific American*. 22 September 2021. 7 November 2022. <https://www.scientificamerican.com/article/agroecology-is-the-solution-to-world-hunger>

12. Shrubsole, Guy. 'How the Extent of Country Farms has Halved in 40 Years.' *Who Owns England*. 8 June 2018. 7 November 2022. <https://whoownsengland.org/2018/06/08/how-the-extent-of-county-farms-has-halved-in-40-years>

13. Marx, Karl. 'The Nationalisation of the Land.' *Marxists*. 7 November 2022. <https://www.marxists.org/archive/marx/works/1872/04/nationalisation-land.htm>

14. Clare, John. 'The Mores.' Poems Against Enclosure. 7 November 2022. <https://la.utexas.edu/users/hcleaver/357k/357kClareEnclosuresTable.pdf>

FLAME

FLAME stands for 'Farming, Land, Agriculture: A Movement For Equality' and is the vibrant youth branch of the LWA. It brings together young people with a passion for food, farming, land and climate. Organising farm visits, skill-sharing and mobilising for marches and political action, FLAME is working to drive change in our food and farming systems

ANON

Trans on the Land

It is interesting how gendered working the land can be. In a space that is so revered, we still find ways to put Western binaries of sex and gender onto the land around us. This first struck me when I read that 'most fruit trees are hermaphrodites' (meaning they carry both 'male' and 'female' reproductive organs), and the possible readings of this scattered themselves across my mind. The queer power of this phrase is wonderful, the fact that the variation of sexual characteristics is a consistent feature throughout all forms of life on this planet. It is a powerful tool for challenging those that tell trans and queer people that we are somehow an anomaly or simply outside of nature, 'unnatural'. I found it affirming in a strange way, to think that other earthly life could have such variety in the organs we categorise as sexual, it might help to counter the rigid binding of sex organs and gender that our society has clung to for the last few centuries.

What does my womanhood mean for those around me and the interactions I have with them? My male co-workers might challenge me as they would a man, or talk over me as they would a woman. The women I work with might be polite and kind, but there is always a distance which it is up to me to bridge, a gap that I have to squeeze myself into to get by. With the performance, my gender reproduces itself like seeds, with all its own energy and capacity for growth contained inside itself. It requires the right conditions to flourish, but its ultimate form will not be altered by the ebbs and flows of opinion.

My hands are strong, hard and dirty, cracked nail polish and soil adorns them. These are the hands of a woman, whose voice rumbles and moves. Just as my hands can rip and heave, they can also brush and nurture. They give new life as they gently tuck seedlings into rich, dark soil.

In the city or countryside my heart is tethered to the ground. As the people around me neglect me, don't understand me or won't try to, I am always kept in good company by small green strawberry plants poking through dead leaves as they rebuild themselves after the frost. Mud caked around my feet and knees takes on new shapes and patterns every day. Standing in blistering wind or bending until my back is sore, slashing back growth and digging until I ugly-sweat and my face is splattered with mud. My humanity, and my womanhood, are the same. They are renewed in struggle and the tranquillity that follows, softness inspired by the rigidity necessary to survive. All praise to the Divine, Lord of all the worlds, dimensions and realities, and the One who gently supports my soul in one palm.

LUKE HARTNACK

Regeneration

This one's for the youth!
Hear my truth.
I am going to make the people of this planet
move.
For you.
I do not want to hand down a broken planet,
so I offer my hands and head to help
and my heart to heal
as I want you to feel
love and grace.
I want you to walk in nature's bounty,
through food forests in sovereignty.
Abundant in seeds to nourish souls as you
engage in a conversation so old we can hear
our ancestors whisper
knowledge and wisdom
about the beauty of the blister
that comes with working the land.
So, I take my stand, in solidarity
for the next generation
I sacrifice superficial temptation
and work towards
the regeneration of humanity

ADAM PAYNE

Keeping the Embers Alive: The Power of Collective Action

Grower and founding member of the Landworkers' Alliance Adam Payne on his journey back to the land and combining farming with activism and social change

Many of my strongest childhood memories come from the land and the communities that live from it. Growing up on a smallholding in North Devon in the 90s, me and my friends would work long summer evenings bringing in hay for neighbours. We would sit and listen to stories of how they used to cultivate the fields around us by horse, growing cereals that were baked into bread and sold at the market in town, or fed to the cows, whose milk made cheeses that were aged in the larder on the cold north wall of the old farmhouse. We would pick apples from the orchard grass, and taste the fresh juice flowing from the press. We would watch lambs slipping into life, and be shown what to look for in a good oak.

But while I was growing up in this world, learning its beauty and finding my heart bound to the land, it was being pulled down around me. I was arriving into a world of small farms being systematically destroyed, already many long decades into the erosion of peasant culture and economy. It was not gone; but the sound of tractors on summer evenings was fading, and where there once were many, only a few remained.

The culture I grew up in never presented farming as a serious option for those not born to farming families. Food and farming were victims of globalisation, and we were taught as teenagers that their time was over and our future was to be found somewhere else. In the years that followed, I began to understand the extent of the damage that the destruction of land-based dreams in the minds of young people like me, had done.

I travelled and spent time in peasant communities around the world. Everywhere, farmers, peasants and landworkers existed in different

economic and political environments, in some places vibrant, organised, and hopeful; in others battered, devalued, and repressed. But everywhere, the centrality of cultural survival was understood as the seed from which the future could grow, the embers that must be kept alive.

Returning home and looking for a meaningful way to live and work, I found my way back to the land and started working in market gardening, combining farming with activism and social change. Working with OrganicLea, a workers co-operative running a market garden and box scheme on the edge of London interwoven with inspiring, community-rooted training, volunteering, and outreach; I was a part of the founding of the Landworkers' Alliance. We had seen the success of the international movement La Via Campesina and were motivated by the potential to build a voice for grassroots farmers, foresters, growers, and Landworkers here in the UK.

Working with the Landworkers' Alliance brought engagement with established unions from around the world. They helped me to understand that the state of the community I grew up in was not predetermined; despite the huge challenges, the decline was not inevitable.

Spending time with organisers in the French union Confédération Paysanne, I learnt that the enviable position of French landworkers was not an inevitable result of the cultural value placed on food in France; but the hard won achievement of generations of organised peasant farmers who articulated, defended, and advanced their culture and economy in the face of huge obstacles. Deepening the cultural value of food was a strategy to defend the land. Now French landworkers can identify as peasants and evoke a range of values around reliability, quality, locality, care, community, and autonomy; which seem to trace an unbroken link back hundreds of years. But this wasn't a historical inevitability, they have fought to defend their identity, to maintain and evolve the policies and social understanding that protect and enhance their livelihoods and culture.

This can seem an overwhelming path to follow here in the UK, where society is deeply infused with the legacies of enclosure, industrialisation, and colonialism. Many of our ancestors were subjected to a systematic and multi-generational campaign to disconnect people from the land, which created precarious workers for industrial centres and traumatised people who became perpetrators of colonialism. The chain of oppression and suffering caused by this has structured the world we live in today. The results of this enforced disconnection are visible in everything, from our education system and food traditions, to our hedgerows and cities.

From this point the path looks long, especially so when the protagonists are farmers, foresters, growers, and landworkers; working long and exhausting days, often isolated and underpaid. But together we can become more than the sum of our parts and see the hope and the power of collective action. Together we see one another and immediately recognise friends and allies, people who know the same world that we do and whose lives share a common thread with ours. Whether we are from the same county or the other side of the world, we have so much shared experience and so much to gain from understanding each other and working together.

But it also goes far beyond land work; we are in a complex web of shared interests with other working people, striving to build economies and social systems that respect the limits of the earth and the rights of all people.

Globalisation of our food and energy systems extracts wealth, nutrients, and carbon; and moves them around the world to points of profit and control. This creates zones of excess and deficiency: extreme wealth and extreme poverty. The decentralised, localised systems at the basis of our vision of food sovereignty circulate finance as they circulate nutrients, carbon, and energy; creating wealth for everyone and ensuring that health and nutrition are the basis of life available to all, and not commodities only for those who can afford them.

We need to put the refracted and broken pieces together to create power. Our diversity is as essential as our unity; every history and identity has a contribution to make, and without embracing all of them we will not be able to achieve the collective evolution that we all need. Our movement has the vision and potential to bring the control of food, energy, and finance back from corporations into the hands of working people, if we can build the strength and unity necessary to take it.

ROZ CORBETT, CLEMENTINE SANDISON,
DONALD MACKINNON AND PATRICK KRAUSE

Reflections on Crofting in Scotland

Roz Corbett and Clementine Sandison from the Landworkers' Alliance, and Donald MacKinnon and Patrick Krause of the Scottish Crofting Federation, share their reflections and aspirations for how crofters and small scale farmers have changed, and continue to change, the landscape in Scotland

Over the last ten years the Landworkers' Alliance's work in Scotland has really developed, with 220 members, staff members supporting our farmer-to-farmer and advocacy work, and the creation of a Scotland manifesto. It has been inspiring to see the movement for small scale farming grow. The Landworkers' Alliance's work runs in parallel with the Scottish Crofting Federation, an organisation dedicated to campaigning for crofters and fighting for the future of crofting.

In this interview, Roz Corbett and Clem Sandison from the Landworkers' Alliance and Donald MacKinnon and Patrick Krause of the Scottish Crofting Federation share their reflections and aspirations for how crofters and small scale farmers have changed, and continue to change, the landscape in Scotland.

RC   So, to start off, what's been the most significant development in Scotland for small farms and crofting over the last ten years?

DM   I think the biggest change is Brexit and the impact that it has had at a policy level. Now, we're in a position where power sits mostly with the Scottish Government, and that is driving a lot of our work and has a massive impact on our members. Whatever you think about Brexit — and we were openly opposed to it as an organisation — this change in how policy is developed and made should be seen as an opportunity. It could also be a threat, depending on what other actors are involved and what priorities the government decides to adopt. But I think for our organisation, it's a fantastic opportunity to put our case forward and make sure that our members and our type of producers are the ones that are supported going forward.

CS   Yes, Brexit, as Donald said. An interesting thing that I've observed is how many more people are trying to get into farming — particularly new entrants, not necessarily young people, often in the middle stages of their lives, looking for a career shift and trying to do it against all the odds. It feels like there are more people wanting to do land-based work; and there has been growth in market gardening and horticulture across Scotland. There is now a real interest in regenerative farming that didn't exist ten years ago; and the idea is filtering through to some more conventional farming communities, with more people transitioning towards agroecological farming methods. Larger scale farmers are seeing the need to reduce inputs purely for cost reasons and are looking to learn from smaller scale farmers and crofters. There is more openness to traditional low input methods.

      I've also seen a growing interest in regenerative grazing methods — where farmers are managing for taller covers, deeper roots, and improved soil biology; with less reliance on fertiliser, greater biodiversity, and the possibility of extending the grazing season.

DM   I think that's right. And I think you are right to highlight rising costs and inflation, which are going to push a lot of producers into adopting these practices simply out of necessity. We simply cannot afford to continue. The impact is on money in your pocket at the end of the day and winter housing costs are enormous for farmers.

RC   The new entrant grant scheme was such a significant enabler for a lot of people, because it meant that the capital was there to get started. It is disappointing that it has ended and I wonder what the impact is on the take up of farming. I know that it's different with crofting though because you still have the Crofting Agricultural Grant Scheme, funding that supports new entrants.

DM   We were quite critical when the new entrant's scheme ended. It is disappointing that the government didn't find a way of continuing the scheme in some form. We also felt that it was fantastic if you got it, but it was maybe a bit too much for some people starting out. There was probably an opportunity for something more tailored, to make the money go further.

PK   The announcement that the Scottish Government made about the climate change emergency is important for crofting and for small scale farming, there is recognition that smaller scale agriculture is more appropriate. In the last decade, there has been a marked change in the attitude of our politicians towards the smaller scale and local food. The pandemic also changed the way people look at local production, it made people really focus on where they get food.

It has been quite a monumental decade really, hasn't it? We've left the EU, so no longer have what could have been protection — there is certainly a better view of small-scale agriculture, small farms, and smallholdings in Europe than exists in the UK. We have been left in a weird void now, where Scotland doesn't seem to be getting itself together with a new agricultural policy. I think we've got our work cut out!

RC   What can the rest of the UK learn from Scotland? We talked a little bit about different legislation that has gone through recently, which many people think of as being quite radical. What changes do you think have been made?

DM   The intention behind the Land Reform Act [Land Reform (Scotland) Act 2016] and the Good Food Nation Act [Good Food Nation (Scotland) Act 2022] and the rhetoric that goes alongside it is all very laudable, and there is a really good conversation happening around these issues. But I think we are seeing a gap between the rhetoric, the ambition, and what's actually happening on the ground. We have seen the process of moving large land holdings into community ownership stall, and there are several different reasons why that has happened. I don't know how you solve that problem; we are missing the radical edge to politics that we need to deliver the change we want to see.

PK   Since this latest phase of the Land Reform Act, I think we are starting to see some momentum in land reform, particularly for small holdings. The idea of them being regulated seems to be getting a lot of agreement and we need to add to this phase of land reform. We probably won't get a better chance.

CS   Yes, I totally agree. I think Land Reform and the Community Empowerment Act [Community Empowerment (Scotland) Act 2015] have a lot of potential and it is early days — there is a lot more work to do. I have not seen much land transferred to community ownership for agroecology, and food and fibre production. That's the missing link. It is difficult to get hold of any prime agricultural land through the Land Reform Act, so it tends to focus more on 'marginal' land. The Act does not have strong enough teeth to actually put really top-quality land into food production for communities. For urban communities, a lot of the Empowerment Act focuses on access to buildings, but it has not resulted in more allotments and growing spaces for food production.

PK   That is interesting. The allotment system is really good, crofting could learn from that process. If you look at allotment rules, they are very clear and fairly straightforward and simple: if you don't use your allotment, you give it up. People know that, and they accept it. For some reason, we have never really managed to instil that same ethos, so now we have a system that is supposed to run like an allotment system, but doesn't because privatisation has slipped in. What people really want for crofting is to have an external arm's length regulator that sits over in the city; but the system management doesn't function.

DM   What we are really talking about are transfers of power and new power to community use, whether that is for allotments, to access land, or to buy a whole estate. That is where we are hitting a stumbling block — deep down there's a reluctance to transfer power to the people.

RC   What do you see happening in the next ten years, do you think there will be that kind of shift?

DM   In the next ten years I'd like to see the people in power recognise the value of small producers; seeing their value in addressing the problems of the climate emergency and biodiversity crisis, as well as food security. I hope that the government will realise that we are an important part of the solution; and that we develop an agricultural support system that recognises and rewards adequately and appropriately, making it fair and possible for people to work. I hope we have a more constructive relationship between all voices to deliver real outcomes. We are polarised right now; there's animosity between different groups, when we all need to be working together.

CS   I'm hoping that we can really strengthen the narrative around land sharing, as opposed to land sparing; and as Donald said, ensure good livelihoods on the land — producing our own food, fuel, and fibre will result in a more resilient society. We have enough land in Scotland for the population to have a fair share, and to produce more of what we need. With conflicts and climate shocks happening around the world leading to grain shortages, we need to look at a more resilient short supply chain model. I hope that we can demonstrate practical examples of

how to live more sustainably and produce the goods we need to have a good life, without intensification of agriculture.

We need to understand that people are part of the ecosystem. Human society must figure out how to restore biodiversity and soil health at the same time as producing our food, not separated from it. We are a part of the land and part of our landscape; and have been culturally for millennia. That is where the Landworkers' Alliance and the Crofting Federation have similar views and goals, and where small scale farmers can really tell their story.

DM   I agree with Clem about the parallels between what both organisations want to see happen. We should be looking at these areas of common ground and building on them.

PK   Yes, I agree. I would add that in the next ten years, we have the chance to see a big shift if we keep the pressure on. Luckily, there are enough people that recognise that we have got to do something. Getting more people doing more ecologically sound things with our land is going to happen.

Epilogue

OLI RODKER

The Landworkers' Alliance Backwards and Forwards

Landworkers' Alliance co-founding member Oli Rodker looks back at the beginnings and shares his vision for the future of the LWA

Sitting here, on the cusp of ten years, we are in a good position to get a view of the LWA both forwards and backwards.

We can look at our faltering beginnings with affectionate hindsight; and we can dream of our future pathways, whether through the gathering clouds of climate crisis, or a regenerating, bountiful landscape. Or maybe both.

When I attended the 2012 gathering at OrganicLea, and joined in the discussion about becoming a UK member of La Via Campesina, I held a number of principles dear. I hope I've kept them with me through ten years on the LWA Co-ordinating Group.

The first is that the political system in this country is as broken as the soil from the very worst of farms. Access to power and politics is fenced off by numerous economic and cultural barriers, and the state will go to great lengths to prevent progressive change. It will require a deep and structural shift to effect the system change that we need. Reform through Parliament is far too constrained; we need a powerful social movement that can work on numerous fronts.

Secondly, this movement should be international in its outlook — the global climate and nature crisis makes this unavoidable. In an age of globalisation, where production is outsourced around the world, and burning — or offsetting — in the Amazon affects us in the UK, international solidarity has to be a defining feature.

Third, it is essential that our bedrock is real life production. We have to be based in the material growing and making of things that people actually need. This is in part because we want to show that we can produce what is needed without the extractive systems of capitalism, and also because we want to be based truly in and with nature. We don't want the regeneration of nature to be simply a temporary career, but to be a whole life!

Fourth, we don't want to create artificial divisions between different types of landwork. We want to get away from the divisive thinking that separates farming from forestry and has split livestock away from arable farming, that divides urban and rural. What holds us together is our work within local ecologies. The divide we are interested in is the one between systems — those that pursue profit at all costs and those that celebrate life.

I doubt that many of us in 2012 could have foreseen that ten years later we would have over 2000 members and a turnover of over a million pounds! And if our first few baby years were small and informal and slowly putting on weight, as we looked to establish our place in the world; then our second phase was a child of Brexit and Covid-19. Brexit meant there would finally be a change to the system of farm subsidy, with all the accompanying changes within government ministries. And Covid meant that suddenly food growing was officially 'essential', and people cared about where their food was coming from. The LWA grew prodigiously.

But now we're wrestling with the next stage. We are bigger and stronger, and maybe both more beautiful and more ugly. Can we remain committed to a horizontal structure as we grow, and how do we make effective our belief in de-centralisation? Can we be home to both more radical and more conservative farmers? Can we provide more services for our members and provide stronger political education and leadership?

We don't know.

At the Oxford Real Farming Conference (ORFC) in early 2023, I quoted the Italian communist Antonio Gramsci who wrote this in prison in 1930:

> The crisis consists in the fact that the old is dying and the new cannot be born; in this interval a great variety of morbid systems appear.[1]

It perfectly describes the state of most farming and land use systems in the UK today. We know a lot about what agroecology could look like here, but we lack the levers to make it widespread. In the meantime, the government sits back and dismally fails to address numerous social and environmental crises, or actively makes them worse.

At ORFC I went on to talk about a type of 'shifting baseline syndrome'; the idea that a decline in environmental quality cannot be seen because it is so gradual, and each generation doesn't know what has previously been lost.

But this shifting baseline syndrome applies in politics too. In this age of individualism, we are losing the knowledge of how to work together effectively and politically. That knowledge has been socially created over hundreds of years, we have a lot of land history to be proud of. It was just, quite often, a long time ago.

The UK is the birthplace of trade unions: 'The Tolpuddle Martyrs', George and James Loveless, James Brine, Thomas Stanfield and his son John, and James Hammett, 'combined' in agreement to fight for better conditions and wages for agricultural workers in 1834. They were deported, but later pardoned and then returned to England a few years later, following widespread popular protest. They formed a union when unions didn't exist, so we know the unthinkable is possible.

We need to work together more effectively to raise our political game. Not just in the LWA, but across organisations, groups, and communities. We know most people can see the sense in a healthy food system that protects nature and is good for the land. We know most of us don't want to get our food from neo-colonial and racist production systems in other countries. But we will need to collaborate to change the world.

So far, we are yet to bring enough people with us, we are yet to create enough political leverage. We are yet to gather enough political power.

Yes, it is hard work to build alliances, to make compromises, to find common ground rather than to see what separates us; but it is essential work. It is what we have to learn to do again. As Raymond Williams said, 'To be truly radical is to make hope possible rather than despair convincing'.[2]

If you are from the more mainstream end of our spectrum, reach out to see that there are systemic problems, which need more than small adjustments at this crucial and critical time. Don't be frightened of change — change is certainly coming.

If you are from the more radical end of our spectrum, reach out to see that not everyone can or needs to agree with your whole analysis. Look for ways to work together, don't let smaller differences deflect from the greater prize.

So, looking out ahead, what do I see?

The LWA creating regional farm hubs that produce food and regenerate the soil, alongside providing well paid jobs and places of mental solace and recovery?

Setting up our own retail systems which give affordable access to healthy local food, no matter what our income bracket?

Educational farms and forests, working closely with schools and building up climate recovery land armies to adapt to a changing landscape?

Empowering communities to take up ecological restoration and land healing work?

LWA foresters working with the housing industry to provide timber for a UK housing resurgence; which sequesters carbon even as it provides warm and healthy homes?

A revival in textile processing as an industry in this country, so that fibre products grown on sheep or in fields can be made into clothes and curtains and carpets without being first exported to China, doused in chemicals, and sent back?

Our members connecting urban and rural livelihoods to build mutual understanding and mutual aid?

Taking direct action to keep out genetic modification and robot/drone-workers in a new wave of Luddite protest?

Helping build a progressive alliance around existing political structures and parties, one that shepherds the adoption of a People's Food Policy and a People's Land Policy into legislation?

Ten years is past, but there is a lot more history to go.

We make the changes we want to see in the world. No-one else will do it for us.

1.   Gramsci, Antonio. *Prison Notebooks* Volume II, Notebook 3, 1930, (2011 edition) SS-34, Past and Present 32-33

2.   Williams, Raymond. *Resources of Hope: Culture, Democracy, Socialism.* London: Verso Books, 1989

JYOTI FERNANDES

A Letter to Future Landworkers

LWA founding member and Campaigns Coordinator and farmer Jyoti Fernandes, shares her message to future landworkers

It is a bright and frozen morning. Our young calf is lying in the sun, the birds are pecking at old seed heads through the ice, we have temporarily found a solution to the burst pipes, and I've just finished peeling carrots and potatoes, uncovered from storage and held plump in their own soil, to put in a stew on the woodstove. It is finally quiet and I have a few moments to think about you, future land workers, and share a few lessons I have learned.

I started this journey, working the land, because of my children. I was pregnant with my firstborn at the peak of youth, when ideals and energy are at their brightest. I had spent time with land-based indigenous communities, watching how they raised their children alongside their daily work; teaching them about how food grows and animals are raised, how the insects and birds, weather and creatures interact. How to gather wild plants, chop wood and preserve harvests. How to cook. How to warm themselves and learn the language of the landscape around them. I wanted my children to learn these things. So, my husband and I decided to become land workers, travelling from farm to farm to learn the skills. Eventually falling in love with our own patch of land.

It has been a hard journey, as farming skills were not passed to us as children. We have had to relearn our skills from old farmers and books and experience as many of you will also have to do. There has been so much joy as we embed ourselves into the intricate web of nature and community. As we gained the skills to nurture orchards bursting with fruits and nuts, to breed flocks where multiple generations of sheep and cows often stand contently licking each other, to fill the larder with tomato passata for the winter. Often the most satisfying moment of the year is when the freshly pressed apple juice is stacked into the apple juice shed, with the heat from the pasteuriser steaming off into the cool autumn air...

These moments of abundance sit alongside moments of despair. We have lost livestock and crops, grown harvests that we couldn't sell, and fought the authorities to stay on our land. Even now, after over 20 years of farming, the dairy cow who had become my daughter's favourite during the Covid-19 lockdown was lost to complications after birthing. I could see my daughters using skills they learned from their years on the farm to nurture the poorly cow, but it didn't work out. We were sad and deflated, yet, through land work we have learned that you must pick yourself up, and that somehow death is a part of the cycle as much as the life of our new fluffy calf.

Lean into the fact that setbacks sit alongside victories, and from each you will learn. This understanding has become ingrained, as much as hope for the future persistently sits alongside fear in my heart. Coming to peace with this feeling builds resilience. I have often thought that every peasant in the world must feel this way.

Years ago, to find that shared understanding, I sought out a network of land workers. I wanted to be a part of a community building something larger than ourselves, so that all those heartbreaks were collectively held and their solutions our common cause. Yet, I have also found that the politics of people — in communities, in relationships — is often the hardest to negotiate as we find our way back to the land. I built the strength to deal with this as I learned; as each part of nature operates in its own unique way, so do people.

If we embrace and build on that diversity, we can build collective strength.

At the first gathering of La Via Campesina that I attended in Indonesia, I sat alongside hundreds of representatives of 200 million peasants, as they welcomed the Landworkers' Alliance into the most diverse and complex, yet grounded, alliance of human beings I had ever encountered. All the feelings I had felt inside myself were channelled into concrete plans for resistance and building that world. I learned that there are many things we cannot do on our own, but when we do them together, we create magic.

Our movement will do all it can to build a strong foundation of knowledge and power for you to draw from. You will need it. The future world you will be working in will be chaotic and volatile. You will face extreme droughts and frozen spells and rain for weeks on end. There may well be economic collapse, hunger, and extinction on a mass scale.

There is a line in the Land manifesto from which I have drawn inspiration for my work over these many years. It says that when capitalism falls and the stars and stripes fade in the West, 'the land will remain.'[1]

Truly, your work on the land will be the most important work there is. To keep the knowledge of the seeds and breeds in living, working memory as our grains of survival. The patterns you recreate will be ones that will see us through. The webs of relationships will be what we use to rise from the ashes.

My daughters, along with many others, created the land workers youth group called FLAME. They've become the flame that warms my heart as I see these young people working together, respectfully and compassionately. I know that we will have the opportunity to rebuild a better system in a design so much better than before. We have the foresight now to bring back nature as part of a just and equitable society; in a new pattern, based on the old knowledge and skills of our ancestors, but more exciting than ever before in history.

Future land workers — you are our legacy and our hope.

We, as land elders, will collectively hold you to go forth to restore our land with grace, agility, wisdom, and above all, love.

1. 'Manifesto.' *The Land Magazine.* <https://thelandmagazine.org.uk/about>

Contributors

Fern Leigh Albert is a photographer and artist whose work is largely concerned with the land and low-impact living. Her photographs have been published in the Financial Times, The Telegraph. She also keeps chickens and grows her own vegetables and is the founder of Farmers' Fayre — a small scale farmers market on Dartmoor.
FERNLEIGHALBERT.COM

Katie Allen is a sheep and cattle farmer and maker based at Great Cotmarsh Farm in Wiltshire. In 2021 she launched her award winning handmade knitwear collection made from fleeces from her flock. She is passionate about connecting more people with the reality that clothes come from farming and a healthy, regenerative model for textiles from the soil up in her podcast, *Ground to Garment*.
LOOPYEWES.CO.UK

Fred Beer is an artist, technologist, creative strategist and occasional grower.

Joya Berrow is a storyteller, photographer and filmmaker.

Mim Black is a founding member of Rhyze Mushrooms Co-op, a radical community mushroom farm and education project based in Edinburgh.
RHYZEMUSHROOMS.SCOT

Black Bark Films is a female led film production collective that evokes authentic and meaningful conversation by working collaboratively with their clients. Topics range from food and farming activism, community engagement, youth empowerment, gender politics, and providing a platform for marginalised voices.
BLACKBARKFILMS.COM

Joanna Blundell is a writer, a TV news producer and worked as the LWA's first press officer from 2021-2. She is curious about the engagement between humans and the natural world and how to communicate that relationship for the benefit of all.
JOANNABLUNDELL.COM

Jackie Bridgen is a tenant smallholder and permaculture designer in Wiltshire. Like most UK citizens, she has never managed to own land, but has succeeded in farming ten rented acres for 20 years.
CHESTNUTSFARM.COM

Chloe Broadfield works on the market garden at Tamarisk Farm in West Dorset. She has an MSc in Sustainable Food and Natural Resources with a research focus on the human-nature relationships and cultural shifts produced through agroecological farming. Chloe is passionate about inspiring others onto the land and offers courses in agroecology.
THEAGROECOLOGICALWAY.COM

Edwin Brooks is an LWA member from Liphook in Hampshire, where he has a small veg farm and helps run a conservation grazing herd. He is also a writer and musician.
THEBURNINGGLASS.CO.UK

Dee Butterly is a small-scale farmer based in Pembrokeshire, Wales. Dee co-founded an agroecological community agriculture scheme and the farm currently sells to wholesale, box-schemes, markets, cafes, restaurants and grocers. Dee is an active member of the Landworkers' Alliance and has been working for the organisation for the past seven years. Dee was also coordinator of 'A People's Food Policy'. In collaboration with 100 UK based organisations in the food sovereignty movement, this process examined and laid out proposals for the practical implementation of the principles of food sovereignty at a national policy level.

Josina Calliste is the Co-Founder of Land in Our Names (LION), a grassroots Black-led collective committed to reparations in Britain by connecting land and climate justice with racial justice.
LANDINOURNAMES.COMMUNITY

Paul Cookson is a green woodworker and founding director of Green Aspirations Scotland CIC, a social enterprise set up in 2013 to reconnect people to their local landscape, mostly through woodland and traditional crafts. It works closely with community groups to deliver training that is appropriate for their resources and needs.
GREENASPIRATIONSSCOTLAND.CO.UK

Roz Corbett is a market gardener, researcher, and organiser. She has been a member of the Landworkers' Alliance Coordinating Group for four years and has previously worked on Scotland policy and campaigns development for the Landworkers' Alliance. Having recently moved to mid-Wales, she still keeps a firm interest in agroecology in Scotland through her PhD research, looking at the Community Right to Buy in Scotland and how that interacts with agroecology transitions.

Olly Craigan is a father and wannabe forester, currently living in North Pembrokeshire. He fell into forestry after finding the solace of woodlands a necessity for staying sane, and now occasionally shouts loudly about the brilliance and potential of trees and UK forestry.

Becky Davies' love of the land around her and appreciation of those who work on and tend to it was fostered as she served as a local priest in rural Welsh communities. Becky's initial training, however, was as an early musician specialising in sung music from around 1100–1750. She loves writing, playing the piano, harp and singing.

Robyn Ellis is based in Manchester and is part of a grassroots permaculture collective called The Gaskell Garden Project. She practices permaculture design methods and philosophies towards how to live consciously in an urban environment.
THEGASKELLGARDENPROJECT.UK

Jyoti Fernandes is an agroecological smallholder with sheep, cows, and orchards based in Dorset. Jyoti is a founding member of the LWA and coordinates the Policy, Lobbying and Campaigning work of the LWA for UK wide campaigns and England specific lobbying on the agricultural transition.

Isabelle Frémeaux (she/her) co-facilitates The Laboratory of Insurrectionary Imagination with Jay Jordan (they/them) bringing artists and activists together to co-design and deploy tools of disobedience. She lives on the zad of Notre-dame-des-landes, where a decades-long struggle defeated an airport project. Their book, *We are 'Nature' Defending Itself: Entangling Art, Activism and Autonomous Zones*, was published by Pluto/Vagabonds/Journal of Aesthetics & Protest, in 2021.
LABO.ZONE

Holly Game is passionate about ecology and pollinators. She is currently studying MSc Biological Recording and Ecological Monitoring, and learning the ways of the beekeeper with Spiritwood Honey on the Pembrokeshire Coast.
SPIRITWOODWALES.CO.UK

Martin Godfrey co-runs Hilltown Organics and Harvest Workers Co-op. He has been involved with the LWA from its beginnings.
HARVESTWORKERSCOOP.ORG.UK

Jack Goodwin is the middleworld incarnation of an eternal dance between three dreamfigures, the angel of discovery and expansion, the angel of the hearth, and a third figure of wyrdness yet to contort to Jack's remembrance. He lives in Devon between the moors and the sea, delights in well-tended food, occasionally writes poems, and is learning to be an uncle.

Robin Grey is a musician, social historian, grower and land rights activist based in Sheffield. He created the show *Three Acres And A Cow, A History of Land Rights and Protest in Folk Song and Story*, really likes kale, and fun fact, also made and maintained the first Landworkers' Alliance website, and designed early LWA propaganda, until they could afford to get someone better to do it!
THREEACRESANDACOW.CO.UK
ROBINGREY.COM

Lucy Grove is a farm hand based in the Shropshire Hills. She is passionate about regenerating the land and in particular the importance of trees in our farmed landscapes.

Ed Hamer is co-founder and manager of Chagford CSA — a seven-acre market garden on Dartmoor. He was a founding member of the LWA and has worked over the years as the LWA's Press and Policy officer.
CHAGFORDCSA.ORG

Samson Hart is a market gardener, land-tender, and writer, based in South Devon. Aside from growing, Samson is co-founder of the wild, radical Jewish diasporist collective, Miknaf Ha'aretz, and a collaborative associate at gentle/radical.

Luke Hartnack is a travelling landworker and Number One Advocate.

Hempen is a not for profit co-operative pioneering the cultivation and production of certified organic hemp food, cosmetics, and CBD. They are led by their values of sustainability, authenticity and the ethical production of both hemp and CBD products, made with love on their farm in Oxfordshire.
HEMPEN.CO.UK

Alex Heffron is a farmer, father, and student who writes about agrarian political economy.

Kai Heron is a Lecturer in Political Ecology at Lancaster University, and a Research Associate at Commonwealth.

Dan Iles worked as a food sovereignty campaigner for Global Justice Now from 2011–2016. He is now based in Wales and works to communicate agroecology projects for the Soil Association.

Jay Jordan (they/them) co-facilitates The Laboratory of Insurrectionary Imagination with Isabelle Frémeaux (she/her). They bring artists and activists together to co-design and deploy tools of disobedience. They live on the zad of Notre-Dame-des-landes, where a decades-long struggle defeated an airport project.
LABO.ZONE

Tom Kemp is a Director and Head Forester at Working Woodlands Cornwall CIC. He has been involved in the LWA's forestry branch since its first national gathering in May 2022.
WORKINGWOODLANDSCORNWALL.COM

Patrick Krause is Chief Executive of the Scottish Crofting Federation (SCF). Having been raised on a smallholding in Bedfordshire and then having worked with smallholders in Malawi, he attended Edinburgh University to gain an MSc in livestock development. He moved from international to Scottish rural development in 2003 when he joined the SCF. He is particularly interested in the challenges in rural development practice, especially regarding the Highlands and Islands, crofting, and smaller-scale food production.
CROFTING.ORG

Rebecca Laughton is the Horticulture Campaigns Coordinator at the Landworkers' Alliance, where she focuses on advocacy and research in the LWA Horticulture Campaign. For 25 years she has combined land work with campaigning, and is the author of *Surviving and Thriving on the Land.*

Julia Lawton and Andy Redfearn are passionate about local food, community, and wellbeing. They continue to develop their unique system of producing food whilst 'Growing Growers' in the heart of the South Downs National Park, and are Project Leaders of Brighton CSA Fork and Dig It CIC.
FORKANDDIGIT.CO.UK

Sam Lee is a British folk singer and traditional music specialist. His debut album, *Ground of Its Own* was shortlisted for the 2012 Mercury Music Award.
SAMLEESONG.CO.UK

Ru Litherland is a Grower and Tutor with OrganicLea.
ORGANICLEA.ORG.UK

Toni Lötter is a mother, writer, poet, organic gardener, natural beekeeper, and earth-materials artist based in East London. She is currently compiling a fabric-record of natural dyes to celebrate overlooked and under-noticed plants on urban streets.

Donald MacKinnon is a crofter from the Isle of Lewis and Chair of the Scottish Crofting Federation. He keeps a flock of sheep and works as a Development Officer for a community landowner. Donald currently sits on the Scottish Government's Agricultural Reform Implementation Oversight Board (ARIOB), providing input on the development of new agricultural policy.

Sky Miller was a trainee with Moor Trees and is an active member of FLAME, the LWA's youth branch. He lives on Dartmoor and in his spare time he likes to go walking and taking photographs on the moor.

Rosanna Morris is a printmaker and illustrator based in Bristol. She has been a longtime collaborator of the LWA and her imagery has featured the organisation's calendars over the last five years. Rosanna's work is underpinned by an active interest in issues around sustainability, the natural world, food sovereignty, and growing.
ROSANNAMORRIS.COM

Morgan Ody is a peasant farmer in Brittany, West France. Morgan is an organic market gardener farming on 1.3 hectares, and sells her produce through a Community Supported Agriculture Scheme (CSA) and at a local farmers market. Morgan has been very active in the French farmers union, La Confédération Paysanne, for many years, working on issues around access to land, land reform, and land redistribution. Morgan is currently the general secretariat of La Via Campesina.
EUROVIA.ORG

Out On The Land (OOTL) is the LWA's LGBTQIA+ member-led organising group. The group brings together Landworkers' Alliance LGBTQIA+ folx to build solidarity, to network and connect, to raise the voices of queer and trans landworkers, and celebrate all the beautiful ways we can be Out On The Land.
LANDWORKERSALLIANCE.ORG.UK/OUT-ON THE-LAND

Adam Payne is a founding member of the Landworkers' Alliance and has been involved ever since. He is a farmer and grower and co-runs Southern Roots Organics, a vegetable farm on 25 acres in Pembrokeshire.
SOUTHERNROOTSORGANICS.ORG

Oli Rodker is a co-founder of the LWA and has been on the Co-ordinating Group since its inception. He is a Director of the Ecological Land Co-operative, a social enterprise setting up affordable opportunities for new entrant farmers. He has lived and worked around land use, woodlands, environmental activism, and climate justice for over thirty years.

Ione Maria Rojas is an artist, food grower and somewhat of a migratory bird. Currently based in Devon, she loves how plants can be portals, connecting us to people and places across the world.
IONEMARIAROJAS.COM

Clementine Sandison is an artist, facilitator, and aspiring urban farmer based in Glasgow; and a Landworkers' Alliance member for eight years. She co-manages a community food forest in the East End of the city and runs projects enabling women to build mutual support networks and access land for ecological farming. She works part-time for the Landworkers' Alliance and Pasture for Life, supporting peer-to-peer learning between farmers, growers, crofters, and foresters in Scotland.

Adam Scarth is a new entrant farmer based in Leeds.

Kate Scott is based in the South of England. She is a shepherdess, milking her lambs to make sheep milk soaps. She works with Pasture for Life and is a member of LWA.
THEDROVERSDAUGHTER.COM

Lauren Simpson is a new entrant landworker in West Wales managing a market garden to provide ingredients for fermented foods business Parc y Dderwen as part of a mixed small off-grid farm under the Welsh 'One Planet Development' planning policy. With a background in documentary film production, Lauren has enjoyed being part of the LWA's book working group whilst working as LWA's Membership Secretary.

Georgie Styles is a freelance audio producer and food anthropologist sharing human stories behind our food. Her work centres food, land, and climate justice; and she produces compelling radio and podcasts for the BBC, Indi's, charities, NGOs, and unions. She strives to help share the voices left unheard.
GEORGIESTYLES.UK

Dr Isobel Talks is a lecturer, researcher and consultant. Currently based at the University of Oxford, she is working with the LWA on a number of different research projects, including the New Entrant Support Scheme pilot which is funded by Defra. Previously she has worked with organisations including Save the Children, Girl Effect, DFID, and Plan International; specialising in education, international development, gender, climate change, agroecology, and new technologies.

Teknopeasant (Conor O'Kane) is a regenerative grower and seed saver in the beautiful Sperrin Mountains, working with community gardens across Ireland. He is also a traditional singer, musician, and occasional songwriter.

Yahel Tsaidi is based in Jerusalem and is passionate about plants, trees and local growing movements.

Umut Vedat is a documentarist interested in working on the topics of discrimination, minority situations, cultural interactions, relief aid, and environmental subjects. He has collaborated with local and international NGOs such as Greenpeace, Heinrich Böll Foundation, La Via Campesina, European Climate Foundation.
UMUTVEDAT.COM

Fritha West manages the Whitelands Project CIC, a woodland based social enterprise in Hampshire. She splits her time between Perth and Petersfield.
THEWHITELANDSPROJECT.CO.UK

Ashley Wheeler is an organic vegetable grower and seed producer based in Devon. He runs Trill Farm Garden with his partner Kate Norman and was a founding member of the Landworkers' Alliance.
TRILLFARMGARDEN.CO.UK

Dee Woods is a Coordinating Group Member and the Food Justice Policy Coordinator at the Landworkers' Alliance. She is a previous BBC Food and Farming Awards winner for her cooking and a music adventurer known for her dance music sets.

Nikki Yoxall is Head of Research at Pasture for Life and is also currently undertaking a PhD in Agroecological Transitions. She has interests in Holistic Management, agroforestry, native breed cattle and connecting folk with their food. Together with her husband, Nikki runs Grampian Graziers, using agroecological principles to manage farm ecosystems whilst producing beef from rare and native breed cattle in the Northeast of Scotland.
GRAMPIANGRAZIERS.CO.UK
PASTUREFORLIFE.ORG

Image Captions

negotiations over the future of UK agriculture policy
and called for equal access to decision making spaces.

P159   Photo: Clementine Sandison
        Farm Hack Scotland was an event for
        agroecological farmers across Scotland to come
        together to share knowledge, skills and take on
        design challenges.

P169   Photo: Fern Leigh Albert
        LWA joining The Big One Climate protest April
        2023 to communicate the link between farming,
        food and climate justice and the role of
        agroecology in a creating a more sustainable future.

About the LWA

The Landworkers' Alliance is a grassroots union of farmers, foresters and land-based workers in the UK. We campaign for the rights of producers and lobby the UK government and devolved nations for policies that support the infrastructure and economic climate central to our livelihoods. We have a growing membership who we work to support by developing agroecology training and solidarity support networks.

We are members of La Via Campesina, an international organisation of over 200 million peasants, small-scale farmers and agricultural workers unions around the world, and work with them to achieve a global vision of agroecology, food sovereignty and sustainable forestry.

LANDWORKERSALLIANCE.ORG.UK